Rhetorical Ways of Thinking

D1810692

Lillie R. Albert

Rhetorical Ways of Thinking

Vygotskian Theory and Mathematical Learning

with Danielle Corea and Vittoria Macadino

 Springer

Lillie R. Albert, Ph.D.
Lynch School of Education
Boston College
Chestnut Hill, MA, USA

Danielle Corea
Center for Faith and Public Life
Fairfield University
Fairfield, CT, USA

Vittoria Macadino
Lynch School of Education
Boston College
Chestnut Hill, MA, USA

ISBN 978-94-007-9650-8   ISBN 978-94-007-4065-5 (eBook)
DOI 10.1007/978-94-007-4065-5
Springer Dordrecht Heidelberg New York London

© Springer Science+Business Media Dordrecht 2012
Softcover reprint of the hardcover 1st edition 2012
This work is subject to copyright. All rights are reserved by the Publisher, whether the whole or part of
the material is concerned, specifically the rights of translation, reprinting, reuse of illustrations, recitation,
broadcasting, reproduction on microfilms or in any other physical way, and transmission or information
storage and retrieval, electronic adaptation, computer software, or by similar or dissimilar methodology
now known or hereafter developed. Exempted from this legal reservation are brief excerpts in connection
with reviews or scholarly analysis or material supplied specifically for the purpose of being entered
and executed on a computer system, for exclusive use by the purchaser of the work. Duplication of this
publication or parts thereof is permitted only under the provisions of the Copyright Law of the
Publisher's location, in its current version, and permission for use must always be obtained from Springer.
Permissions for use may be obtained through RightsLink at the Copyright Clearance Center. Violations
are liable to prosecution under the respective Copyright Law.
The use of general descriptive names, registered names, trademarks, service marks, etc. in this publication
does not imply, even in the absence of a specific statement, that such names are exempt from the relevant
protective laws and regulations and therefore free for general use.
While the advice and information in this book are believed to be true and accurate at the date of publication,
neither the authors nor the editors nor the publisher can accept any legal responsibility for any errors or
omissions that may be made. The publisher makes no warranty, express or implied, with respect to the
material contained herein.

Springer is part of Springer Science+Business Media (www.springer.com)

To our families, mentors, and friends

Preface

As Vygotskian theory becomes more broadly known and applied to mathematical education research, the sociocultural context has philosophically influenced how we think as well as what we think about pedagogical content knowledge. Vygotsky (1978) contended that research should result in a dynamic analysis in which "the complex reaction must be studied as a living process, not as an object" (p. 69). From a historic perspective, which concentrates on the origin of the experience and its developmental history, researchers must search for methods that will assist in developing our understanding of human activities. Specifically, it is imperative to study the complex nature of mathematical learning processes using a method of dynamic analysis to understand their influence on pedagogical approaches. The aim of this book is to present the main outlines of sociocultural historic theory, drawing as much as possible on Vygotsky's major constructs; however, it is key that they are situated in non-complicated ways, so that they may be followed by those who are not researchers.

The constructs covered in this book to a large extent provide a broad examination of sociocultural-historic theory as it relates to mathematical pedagogical knowledge. An argument highlighted in this book is that the major assumptions of sociocultural-historic theory are essential to understanding the theory's application to mathematical pedagogy. As an innovation, since this has not been done much in the field of mathematics education, an aspiration is to demonstrate how to in effect merge the theory with practice. In particular, the empirical studies presented in this book illustrate ways to study mathematical thinking and learning over time. What's more, the studies provide a point of view regarding the importance of understanding the origin or history of teachers' mathematical knowledge; thus, these studies apply a mixed methodological framework to investigate and to make sense of teacher-generated images and drawings and how collaborative contexts make possible effective scaffolding or socially shared thinking and learning. Therefore, this book draws attention to methods framed within sociocultural historic theory to inform mathematics researchers and teacher educators, professional development providers, and policymakers. This work is appropriate for prospective and practicing teachers as the content elements used to model various theoretical constructs of sociocultural theory are practical classroom examples from various grade levels (Kindergarten to Twelfth Grade). Furthermore, it might be useful for advanced undergraduate and graduate students who are interested in critically examining mathematics teaching and learning through sociocultural historic theory lens.

This book demonstrates ways of representing the self-regulation of mathematical knowledge and experiences and examines how "knowing" is the authentication of one's thinking articulated through underlying epistemological formations. It effectively models how mathematical teaching and learning might be informed by, and contributes to our understanding of the importance of positioning thinking and learning in sociocultural contexts. "Only the objectification of the inner process guarantees access to specific forms of higher [performance] as opposed to subordinate forms" (Vygotsky, 1978, p. 75). Using Vygotsky's methods may be of assistance in encouraging us to pay closer attention not only to the product of performance as a final consequence, on the contrary, we may be encouraged to study profoundly the underlying processes or experiences that help learners of all ages "grow into the intellectual life of those around them" (p. 88).

Lillie R. Albert

Reference

Vygotsky, L. S. (1978). *Mind in Society: The development of higher psychological processes.* Cambridge, MA: Harvard University Press.

Acknowledgements

A little over 10 years ago after returning from the *International Vygotsky Society's Summer Conference on Psychological Development and the Zone of Proximal Development*, I began to develop research studies that would assist my understanding of how sociocultural contexts affect mathematical thinking and learning. This book represents some of my understandings, especially the idea that mathematical thought takes place outside and beyond the mind – returns to the mind – and then it evolves in another pathway built on fresh, yet different, thoughts granted through sociocultural experiences. Therefore, I appreciate the time I spent with my colleagues in Moscow, Russia. I especially like to thank my colleagues Michael Schiro, Boston College, and John McAdam, Marist College, for encouraging me to attend the conference.

I would like to thank the prospective and practicing teachers, my undergraduate and graduate students, and my entire group of graduate research assistants and undergraduate research fellows who collaborated with me to complete this work. In particular, I am grateful to the following: Christopher Bowen, Danielle Corea, Michael Egan, Abraham Kim, Eun Sil Kim, Rina Kim, Andrea LaGala, Esther Lee, Vittoria Macadino, Gail Mayotte, Katherine McKee, Cynthia Phelan, Jansen Po, Kathleen Rhoades, Sheila Sohn, Jessica Tansey, Karen Terrell, Yan Zhao, and Peiyun Zhou.

Finally, my sincere appreciation goes to a mentor and collaborator, Professor Jong-Soo Bae of Seoul National University of Education, Seoul, South Korea.

Contents

Chapter 1
Introduction

A "brilliant and charismatic thinker, speaker, and mentor" is an accurate depiction of the psychological theorist Lev Vygotsky, but even this blazon falls short in describing his tremendous abilities and contributions to society; without Vygotsky, the face of modern psychology would not be the same (Newman & Holzman, 1993, p. 5). Even before he became an exceptional developmental psychologist, Vygotsky was a cultural theorist and teacher. After his focus shifted to psychology, it was through his theoretical research and experimentation that Vygotsky strongly influenced the historical transition to psychology as a human science (Newman & Holzman). He also played a crucial role in the fields of human development and education. Vygotsky's belief that "humans are significantly influenced by the sociocultural, or social and historical, context that mediates their experience" created many new initiatives for teaching that are employed in modern-day Western education (Samaras, 2002, p. xxi). He always viewed education as a way to improve one's potential; this, in turn, led him to pursue the development of a complete pedagogical theory focused on discerning concrete educational methods to best achieve this goal. Early implementation of Vygotsky's instructional methods in the United States dealt with literacy learning and childhood growth and development, yet it was not until the past two decades that the application of his theory was extended to include mathematical pedagogy (Bruner, 1987; Samaras, 2002).

Why Rhetorical Thinking?

This book, *Rhetorical Ways of Thinking*, focuses on how the co-construction of learning models the interpretation of a mathematical situation. It is a comprehensive examination of the role of sociocultural-historical theory developed by Vygotsky. The aim of this book is to put forward the supposition that the major assumptions of sociocultural-historic theory are essential to understanding the theory's application

L.R. Albert, *Rhetorical Ways of Thinking: Vygotskian Theory and Mathematical Learning*, DOI 10.1007/978-94-007-4065-5_1,
© Springer Science+Business Media Dordrecht 2012

to mathematical pedagogy, which explores issues relevant to learning and teaching mathematics-in-context, thus providing a valuable practical tool for general mathematics education research. *Rhetorical Ways of Thinking* for all intents and purposes is a way of modeling self-regulation of and self-reflection on mathematical knowledge and experiences. Furthermore, our argument is that this way of knowing is the substantiation of one's thinking expressed through underlying epistemological formations and experiences. The goal is not just to influence the reader's thinking and understanding of sociocultural-historic, but rather to illustrate how to effectively merge sociocultural-historic theory constructs with practical applications of mathematics pedagogy.

This inquiry into Vygotskian theory is conceptually grounded in established methods and practices in the field of education and is theoretically driven, presenting some of the major assumptions of Vygotsky's sociocultural historic theory that are fundamental to understanding the theory's practical application to mathematical pedagogy. It offers an in-depth overview with references to the work and research from which this work is derived. This inquiry involved reviewing over 200 empirical and theoretical studies in which the major framework was sociocultural theory. The areas of review and analysis were extensive and have expanded over 10 years, ranging from work in psychology, philosophy, anthropology, education, and cognitive and learning science. Furthermore, more than 150 books on this subject, including the collected volumes of L. S. Vygotsky, were examined to develop a more adequate picture of how the theory relates to mathematics pedagogy.

The context for this inquiry was my attendance at Vygotsky Psychology Institute and observations of practical applications of Vygotsky's theories at the Golden Key Summer School, which are supported by scholars and graduate students of Russian State University for the Humanities. The Institute focus was higher psychological education, which carried out scientific work by theorists and their students. The Golden Key school environments offered participants the opportunity to take part in presentations and classroom activities, learning first-hand about how a Vygotskian method is practiced in education and to develop an understanding of how the zone of proximal development is interpreted.

Overview of Chapters

The plan for this book is as follows. Chapter 1 provides an introduction to the book, including a brief overview of each chapter. Chapter 2 serves as a primer to Vygotsky's sociocultural historic theory. Chapter 3 is an empirical study examining prospective teachers' perception of mathematics teaching and learning. Chapter 4 explores how scaffolded instruction assists practicing teachers in developing their pedagogical content knowledge. Chapter 5 is the contextualization of the theory to practice.

Chapter 2 is the heart of this book. Major data sources for this chapter include journals and logs of notes from my readings and observations made during my participation in the Golden Key Summer School activities. In this chapter, we present some of the major assumptions of Vygotsky's sociocultural historic theory that are fundamental to understanding the theory's practical application to mathematical pedagogy. The idea is to provide some sense of how the theory relates to the mathematical teaching and learning. It was Vygotsky's belief that the historical feature of intellectual development merges into the cultural one, and the tools learned thoroughly by us to master tasks in our environment and to control our own behavior were invented and perfected in the long course of our social history.

Through a series of drawings, narratives, and focus group interviews, the research presented in Chap. 3 investigates prospective teachers' perceptions of teaching and learning mathematics at the elementary level. They were asked to illustrate mathematics teaching situations of their past, present, and idealized future classrooms. A *theoretical framework* guided data collection and analysis, which characterized three unique experiences: past reflection of mathematical experiences as a student (*Remembering*), current experience as a prospective teacher (*Apprenticing*), and the future image of the student in the role of the practicing teacher (*Actualizing*). Analysis of the drawings applied a method developed by Haney, Russell, Cengiz, and Fierros (1998), which included creating a coding scheme; then, applying Kappa estimates to measure for inter-rater reliability. These drawings and the focus group interviews revealed that the prospective teachers' perceptions of old, new, and idealized mathematical teaching experiences demonstrated an evolving awareness of pedagogy and curricula that constitute effective mathematics instruction.

In Chap. 4, we examine research concerning scaffolded instruction that promotes collaborative learning of mathematics content. Sociocultural historic theories have been introduced from associated fields to suggest that cognition and learning takes place at individual and group levels. The concept of a scaffolded instruction model, the *Field of Social Interaction*, was developed to explain how multiple individuals share meanings and understandings of mathematical content. Applying a mixed-methods design, this study explored the concept of shared meanings through the language of scaffolding among practicing middle school mathematics teachers. Mathematics instruction is offered during a professional development program designed to increase mathematical content knowledge. Findings suggest that when learners work in collaborative situations, scaffolded instruction may provide opportunities for those learners to be the knowledgeable others, especially when linked to intentional and deliberate pedagogy. This research provides a fresh perspective on the role of learning and understanding mathematical content within a collaborative context in which teachers' metacognitive processes evolve and influence their role as teachers of mathematics.

In Chap. 5, we discuss what we have learned from the merging of Vygotskian theory with practice. For example, we have learned that the intellectual development of individuals arises from one's culture, which includes the thought, language, and reasoning processes emerging from social interactions with others to create a joint knowledge of the culture. Intellectual abilities and processes were studied to discern how the historical sequence of events that produced these abilities related to one's culture; therefore, culture is essential to intellectual development. What one thinks about knowledge and the process of thinking itself is acquired from one's culture through the use of tools of intellectual thought. These aspects of sociocultural historic theory are applicable to studying intellectual development in general, especially in children's learning and development, but should not be limited to children alone. The research presented and discussed in this book makes it clear that sociocultural contexts should include understanding learning and development across the life span, and when focusing on teaching and learning, particular emphasis should include the study of mathematical learning and teaching. The empirical studies exemplify the practical applications of Vygotskian theory in which the goal is to establish ways to conduct research using Vygotsky's methods for studying mathematics thinking and learning over time.

References

Bruner, J. (1987). Prologue. In *L. Vygotsky, the collected works of L. S. Vygotsky* (M. Cole, S. Scribner, V. John-Steiner, & E. Souberman, Trans.). Cambridge, MA: Harvard University Press.

Haney, W., Russell, M., Cengiz, G., & Fierros, E. (1998). Drawing on education: Using student drawings to promote middle school improvement. *School in the Middle: Theory and Practice, 6*(5), 38–43.

Newman, F., & Holzman, L. (1993). *Lev Vygotsky: Revolutionary scientist.* New York: Routledge.

Samaras, A. P. (2002). *Self-study for teacher educators: Crafting a pedagogy for educational change.* New York: Peter Lang Publishing Inc.

Chapter 2
Vygotsky's Sociocultural Historic Theory, A Primer

Ultimately, only life educates, and the deeper that life, the real world, burrows into the school, the more dynamic and the more robust will be the educational process. That the school has been locked away and walled in as if by a tall fence from life itself has been its greatest failing. Education is just as meaningless outside the real world as is a fire without oxygen, or as is breathing in a vacuum (Vygotsky, 1997c, *Educational Psychology*, p. 345).

Introduction

One of the most dynamic perspectives that has captivated the education field is sociocultural historic theory. It is the name given to the Vygotskian approach, which emphasizes the cultural context of learning and development. A basic premise of this approach is that the origins of higher mental functions are uniquely human and are found in our social relations with the external world. An essential aspect of this idea, noted Vygotsky, is that we are not just products of our environment but we are also active agents in creating that environment. According to Luria (1979), Vygotsky referred to his approach as "cultural and historical" to reflect new ways of studying learning and development and as a way to distinguish human intellectual developments from those of lower animals. The cultural aspect of Vygotskian theory notes "the socially structured ways in which society organizes the kinds of tasks that the growing [human being] faces and the kinds of tools, both mental and physical, that a [human being] is provided to master those tasks" (Luria, p. 44).

Tools function as the conductor of human influence on the objective of an activity, resulting in changes in that objective. For example, in a mathematics class if my objective is for students to solve a problem using the quadratic equation, then the tool might be the graphing calculator. That tool will influence the objective, as the problem will require less time to complete with the aid of the calculator, depending on the nature and structure of the problem. Tools are *externally* oriented, serving as a

L.R. Albert, *Rhetorical Ways of Thinking: Vygotskian Theory and Mathematical Learning*, DOI 10.1007/978-94-007-4065-5_2, © Springer Science+Business Media Dordrecht 2012 means by which human external activity aims to conquer nature (Vygotsky, 1978, p. 55). A notable example of a tool invented by humans is language, which plays a major role in the "organization and development of thought process" (p. 44). It was Vygotsky's belief that the historical feature of intellectual development merges into the cultural one, and the tools learned thoroughly by us to master tasks in our environment and to control our own behavior were invented and perfected in the long course of our social history. "Language carries within it the generalized concepts that are the storehouse of human knowledge. Special cultural instruments like writing and [mathematics] enormously expanded [our] powers, making the wisdom of the past analyzed in the present and perfectible in the future" (Luria, 1979, p. 45).

In this chapter, I present some of the major assumptions of Vygotsky's sociocultural historic theory that are fundamental to understanding the theory's practical application to mathematical pedagogy. I offer only a brief outline with references to the work and research from which my outline is derived. The idea is to provide some sense of how the theory relates to mathematics teaching and learning. The analysis and assumptions presented here serve as a primer and it is suggested that the reader must consult these works to develop a comprehensive picture.

Tools of Human Intensification

At the heart of Vygotsky's theory is the notion of mediation, or semiotics. Semiotics is the study of the tools and signs that are elements of communicative processes or systems, which allows for qualitative changes in intellectual development (sociocultural historic). The interlocking concepts of mediated activity and tools of the mind (psychological tools) are products of our cultural history. Vygotsky provided a number of examples of semiotic means, including language, various systems of counting, mnemonic methods, algebraic symbol systems, works of art, writing, schemes, diagrams, maps, and drawings, which are extended to technological tools used in mathematics classrooms such as graphing calculators, software programs, computers, and Smartboards.

Intellectual development then rests on the internalization or mastery of the tools of one's culture. That is, tools emerge and change as culture develops and changes. For example, in mathematics education, we can consider technological tools as instructional tools that can be used in various ways. In the 1600s, quill pens and slates were used to teach students how to write and cipher, in 1901 Maria Montessori's kinesthetic method offered a multiplicity of manipulatives from which students could learn, in 1956, Bloom's taxonomy of educational objectives in three domains, cognitive, affective, and psychomotor was published, and, in 1957, the emergence of programmed instructional materials based on Skinner's behaviorism was implemented in elementary schools (Reiser, 2001a; Shrock, 1995). A few decades later, with the arrival of personal computers, e.g., the Apple IIe, many mathematics educators embraced computer-assisted instruction, which reached its peak in the 1980s, leading to other forms of computer based technologies such as video discs, CD-ROMS, multimedia, digital presentations, interactive video, and the Internet (Heinich, Molenda, Russell, & Smaldino, 1999; Reiser, 2001b; Shrock, 1995). Today, the Internet and Web 2.0 offer numerous forms of reality that let teachers and students experience learning anytime and anywhere. One only needs to visit sites such as YouTube or Twitter to interact virtually via live video and audio. All of these technological tools, which are essential to the appropriation and acquisition of knowledge through activities or tasks are also tools used and mastered by individuals of a society, promoting human intensification or development. As the tools of our culture emerged and changed, we changed as well. As society changes and begins to rely on advanced cultural tools, classroom use of these tools also changes how mathematics is taught.

Mediation Through Signs

The notion discussed in the previous section brings us back to the issue of mediation and psychological tools in a mathematics classroom, especially since both become important in the context of social activities and organizations leading to specific patterns of behavior, which connects all students to the learning community and to themselves. From a Vygotskian stance, signs have specific meanings that develop overtime by humans to fulfill the role of psychological tools (Kozulin, 1998). Simply stated, mediation is the use of a sign or symbol to represent a specific behavior or another object in one's surroundings. The source of mediation can be found in material tools (base-ten-blocks or algebra tiles), in the system of symbols (our algebraic system), or in the behavior of another human being (assisted learning by a teacher or a more knowledgeable student).

Mediation is an essential aspect of the process of *concept formation*. This complex process requires interactions between all basic intellectual functions. An integral part of the process of concept formation involves the use of signs, which are "artificially created stimuli whose purpose is to stimulate behavior, to form new reflex connections in the human brain" (Ghassemzadeh, 2005, p. 289). Because we cannot govern our own behavior directly, we use a sign system consisting of mnemonic devices, speech, and writing, in order to mediate and therefore control our behavior indirectly. Signs serve as a means by which to direct and control the course of our mental operations and, subsequently, to guide us toward the solution to a problem. Initially, signs act as a means of creating external social connections with others. That is, they serve an interpsychological purpose; whereas, later signs become a means of influencing one's own thought process. The social aspect serves as the starting point of semiotics, because the individual eventually transfers or internalizes the social relation into one's own self-regulation (Ghassemzadeh). "A sign is always a means used for social purposes, a means of influencing others, and only later become a means of influence oneself" (Ghassemzadeh, p. 550). Learning to direct one's own intellectual processes using signs eventually leads to the development of higher mental functions and internalized abstract thought (Holborow, 2006).

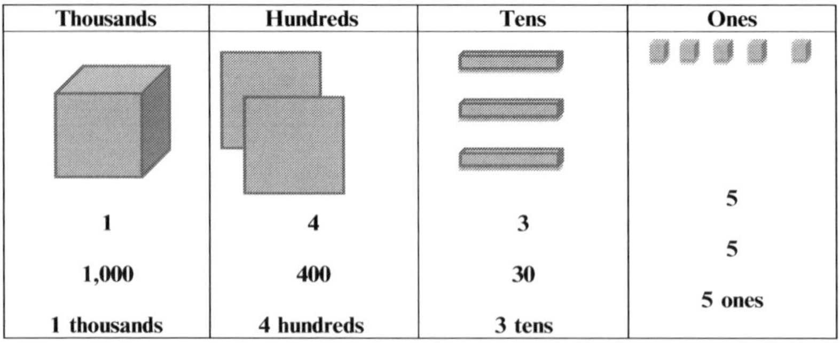

Thousands	Hundreds	Tens	Ones
1	4	3	5
1,000	400	30	5
1 thousands	4 hundreds	3 tens	5 ones

Fig. 2.1 Base-ten-blocks

According to Vygotsky, in order to mediate behavior, a sign, or second order stimulus, functions as an intermediate link between the stimulus and response, forming a new relation between the two. While the sign operates on the individual, the individual must be actively engaged in establishing this link; thus, signs are self-regulated and have both communicative and intellectual functions. Kondratov (1969) classifies signs in three ways. The first is the natural sign, which is based on a sequential or causal connection such as that of smoke indicating fire. Second is the iconic or copy sign, which is based on resemblance. And last are symbolic signs, which are arbitrary. They require the active participation of the subject to make the signifying connection (Ghassemzadeh, 2005). Symbolic signals can include signals of communication or conventional signs such as waving a hand to call for help or to greet someone.

Thus far, Kondratov's (1969) different types of signs have only been classified in terms of their mediation in language and linguistic concept formation. However, this classification of signs can also be applied to the realm of mediation in mathematics concept formation. Examples of natural signs in mathematics education include Base-ten-Blocks and Algebra tiles. Through Base-ten-Blocks (Fig. 2.1), students sequentially begin to understand the concept of counting, place value, and basic operations. Whenever students see a unit cube, they understand that it represents one unit (1), whenever they see a long or rod, they know that it represents 10 units (10), and similarly, they realize that a flat represents 100 units (100). Algebra Tiles are concrete models of variables and integers that help students understand and explore algebraic concepts. These tiles are based on an area model as illustrated in Fig. 2.2. The flat represents x^2, the long represents x, and the unit represents 1.

The graph of a quadratic function is a mathematical example of an iconic sign. Like a picture of a tree, which is an iconic sign to help children conceptualize an understanding of a 'real' tree, the graph of the quadratic function is a two dimensional representation of the parabola formed by a quadratic function. In the elementary grades, iconic signs in mathematics are readily used in the form of pictures or other visual representations. Vygotsky's understanding of 'sign' is most comparable to the symbolic sign. Vygotsky maintained that the learner is enabled to interpret the information or action provided by the sign. For example, in mathematics, the symbol θ,

Fig. 2.2 Algebra tiles

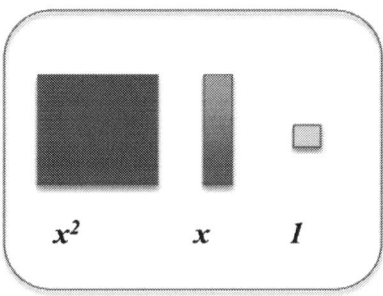

$$x^2 \qquad x \qquad 1$$

Table 2.1 Mediation through sign systems

Type of signs	Natural signs	Iconic/copy signs	Symbolic signs[a]
Description	Based on sequential or casual connection	Based on resemblance	Arbitrary, require active presence of an "interpretant" to make signifying connection
Examples	Smoke – sign of fire Base-ten-blocks Algebra tiles	Picture of a tree The graph of a quadratic function	Signs of communication or conventional signs- string tied to a finger, written note, piece of poetry A graph of a set of data, e.g., population growth

Note. Adapted from the work of Kondratov (1969)

[a]Vygotsky's use of signs and sign systems is referred to as the symbolic sign

"theta" may not be anything more than a Greek letter to a non-mathematician. However, after being taught in pre-calculus that, in mathematics, this symbol always represents the measure of an angle, students will begin to interpret this symbol as the measure of an angle. Another similar example would be pi (Π). Descriptions and examples of these different types of signs are compiled in Table 2.1.

As we have seen thus far, Vygotsky's psychological and theoretical ideologies about the role of tools and signs in intellectual development can provide an explanation of students' learning and understanding of mathematics skills and concepts. Vygotsky's work along this line also suggests that learners of mathematics have the aptitude "to act purposefully according to socially meaningful goals and with the help of socially developed tools, thus overcoming the dictates and constraints of nature and environment" (Stetsenko, 2004, p. 504). Vygotsky and Lauria (1994) suggest that to act purposefully "is in direct functional dependence on the use of signs" (p. 166). Furthermore, an important consideration in understanding the role of signs in developing the individual capability to learn and understand mathematics concepts, is not to view this role in isolation, but rather to view it as a dynamic as well as a practical sociohistorical process, which helps in illuminating the fundamental nature and development of psychological functions.

The Development of Higher Psychological Functions

As explicated in the introduction to this chapter, the origin of the development of mental functioning is integral to understanding Vygotsky's sociocultural historic theory. Vygotsky (1978) argued that the only way to study higher psychological functions is to understand their beginnings, giving consideration to how development of higher forms is determined. In elaborating on this argument, Vygotsky explained that development must be studied historically. *"To study something historically means to study it in the process of change"* (emphasis in original, p. 65). Vygotsky offered these examples:

> We might study the development of memorizing in children by making available to them new means for solving the given task and then observing the degree and character of their problem-solving efforts. We might use this method to study how children organize their active attention with the aid of external means. We might trace the development of arithmetic skill in young children by making them manipulate objects and apply methods either suggested to them or "invented" by them. What is crucial is that in all these cases we must adhere to one principle. *We study not only the final effect of the operation, but its specific psychological structure* (emphasis in original, p. 74).

Therefore, in understanding the origins of these tasks, we concentrate on the very process of how learners complete these tasks, and in consequence, we are able to "discover the inner structure and development of the higher psychological processes" (p. 74).

Distinguishing Between Lower and Higher Functioning

According to Vygotsky, the development of human mental functions or psychological functions follows a process of transformation from the form of lower mental functions to the form of higher mental functions. Vygotsky defined lower or elementary mental functions (LMF) as biologically programmed, natural behaviors, or immediate responses to stimuli that occur automatically, without mental thought or consciousness (Ratner, 2004). Lower mental functions are the simple, primitive, and independent responses that are developed naturally in both humans and higher animals. Some examples of these functions are sensation, reactive attention, and spontaneous or associative memory.

By contrast, higher mental/psychological functions (HMFs) are deliberate, mediated, and internalized behaviors. They are built upon lower mental functions in a culturally specific way. Human consciousness serves as a "mental space" of psychological phenomena that includes perception, memory, thinking, language, emotions, etc., which mediates between a stimulus and response (Ratner, 2004). These cognitive phenomena monitor new stimuli and produce a response that is intentionally and appropriately applied. HMFs require a completely different level of thinking than that of higher animals; thus, through evolution and the development of civilizations, humans are the only beings that have acquired higher mental functioning such as mediated perception, focused attention, deliberate memory, and logical thinking.

Vygotsky differentiates LMFs and HMFs in four distinct categories through examining origins, structure, and ways of functioning and relation to other mental functions (Subbotsky, 1996). LMFs are genetically innate, unmediated in terms of construction, spontaneous in terms of functioning, and isolated as "individual mental units." By contrast, HMFs are socially achieved, mediated by social meanings, voluntarily controlled, and united in systems with other functions as opposed to existing as individual units (Subbotsky).

It is Vygotsky's belief that "the most fundamental qualitative change over the life-span ... is from lower, elementary processes to higher, conscious, psychological processes" (Ratner, 2004, p. 401). It is the transition from "direct, innate, natural forms and methods of behavior to mediated, artificial, mental functions that develop in the process of cultural development" (Vygotsky, 1998, p. 168). Vygotsky outlines four major criteria that distinguish higher from lower psychological functions (Wertsch, 1985). We will examine these differences in the context of a mathematics example: children learning to count. At the age of 3 or 4, a child is usually able to count sequentially from one to ten, but at this point, the ability to count exhibits a simple mastery of memorization, a lower mental function. As the child learns in kindergarten that numbers represent quantity, the child begins to develop a conceptual understanding of numbers, leading to the development of higher mental functions. Specific classroom instruction is required for this abstract understanding and development to occur.

The first general criterion that distinguishes higher from lower mental processes is that control shifts from the environment to the individual. In other words, higher mental processes are governed by self-regulation whereas lower mental processes are subject to the control of the environment. In our counting example, the young child learns the order of the numbers from one to ten by imitating others (i.e. parents or a character on television). Without an understanding of the meaning of the numbers, the child could just as easily count "1, 3, 2, 4, 5…10" if this is the order that is constantly reinforced. Consequently, this process of counting is subject to the control of the environment. However, when the child understands that concept of quantity and the sequential purpose of counting, she realizes that order is important. For example, she recognizes that two precedes three because two cookies are less than three cookies. Thus, the child, at this point in her learning, is able to regulate her own understanding of number.

The second distinguishing factor is that conscious realization or intellectualization is present in higher but not lower mental functions. This relates to the first distinguishing factor. For example, when a young child counts to ten to imitate a character on a TV show, she may do so without an understanding of quantity (i.e. without conscious realization), whereas after the child has learned the concept of quantity, she may count with a purpose, such as to determine how many pennies she has, or how many goldfish she wants.

The third distinguishing criterion is that higher mental processes have social origins and social nature whereas elementary functions do not. Vygotsky considers society, not nature, the determining factor in human behavior (Wertsch, 1985). Thus, for Vygotsky, an essential part of understanding an individual's development consists of comprehending how social interactions, particularly in small groups or dyads, lead to the development of higher mental functioning (Wertsch). An example relating to counting involves a child who is able to count meaningfully (sequentially) from one to ten, but who struggles with counting to 20. With the assistance of another student who is capable of counting meaningfully from 1 to 20 and under the guidance of the teacher, the child successfully completes the task. Such assistance can involve concrete models or manipulatives to help the child visualize while counting, or it can simply entail the children counting together orally. Either way, the social interactions between the two children eventually facilitate meaningful counting for the less experienced learner. In due course, the child is able to count to 20 by himself or herself (without the assistance of his peer), developing this HMF from the outside in. This example supports Vygotsky's (1960) conclusion that the "transition from a social influence external to the individual to a social influence internal to the individual" is of central importance to the development of higher mental functioning (p. 116). An important distinction, noted Vygotsky (1978), is that higher mental functions have their social roots and historical development in meaningful interactions, concluding that an *interpsychological process* (with others, outside) changes into an *intrapsychological process* (with self, inside).

> Every function in the child's cultural development appears twice: first, on the social level, and later, on the individual level; first, between people (*interpsychological*), and then inside the child (*intrapsychological*). This applies equally to voluntary attention, to logical memory, and to the formation of concepts. All the higher functions originate as actual relations between human individuals (Vygotsky, 1978, p. 57).

The last distinguishing factor is that higher mental processes are functions of mediated activity, which is not the case for lower mental processes. Vygotsky saw the first three criteria as evidence that "presupposed the existence of psychological tools or signs, that can be used to control one's own and others' activity" through mediation (Wertsch, 1985, pp. 26–27). Mediation involves an object or symbol (i.e. material or psychological tool) used to represent a particular behavior or another object in the environment (Kozulin, 1990). For example, in mathematics, counting is a mediated activity; the number 'three' mediates perception of quantity. Some other mediated activities include writing and drawing, and some mediated processes include concept formation, memorization, and problem solving. It is through mediation that an interpersonal process is transformed into an intrapersonal one, which is fundamental to the formation of HMFs (Wertsch).

Although some LMFs are prerequisites for HMFs, such as involuntary to voluntary memory, HMFs are not merely a continuation or combination of elementary functions. Instead, "they are a qualitatively new mental formation that develops according to completely special laws and is subject to completely different patterns" (Vygotsky, 1998, p. 34). Let us revisit the counting example to clarify this idea. First, memorization of the numbers from one to ten (LMF) is a prerequisite for conceptualization of quantity (HMF). However, there are many aspects involved in developing this abstract understanding of numbers; in addition to reciting numbers sequentially, students must also recognize them numerically, and associate them with specific quantities of concrete materials. With formal instruction, students are able to combine all of these aspects into a unified concept of quantity; consequently, a "new mental formation" is developed. For example, for students to fully understand the concept of "one," they must realize that it can be applied universally – one apple, one person, one duck, and so on. They must also recognize that the numeral "1" represents this common "quantitative set" or classification. In addition, students must identify that sequentially, the counting numbers begin with "one." All of these aspects contribute to the development of a new mental formation or conceptual framework for the quantity of "one." In this example, we have examined the complexities of Vygotsky's (1997a) notion of the formation of HMFs, including how their development is influenced by "completely different patterns" than those that govern LMFs.

In summary, the geneses of lower mental functions are involuntary unmediated formations, mostly individual; whereas higher mental functions are voluntary, controlled formations obtained and mediated by social meanings, generally linked to an expansive system of functions. The development of higher cognitive functioning involves spreading out mathematics instruction and learning experiences from the outside in, from the interpsychological to the intrapsychological, from interaction with others to self, from other-regularization to self-regularization. Then the cognitive processes and capabilities are formed and built up in part by increasing learners' support, intensifying the competence of learners to reflect on experiences and learn from them. To broaden meaningful social interactions, learners must familiarize themselves with the concept being studied, and decide on a mode of understanding. These aspects help make possible the acquiring of *"higher mental functions* – deliberate, symbol-mediated behaviors that may take different forms dependent on the specific cultural context" (Bodrova & Leong, 2007, p. 9). A key Vygotskian stance is that the transformation of cognitive processes (interpsychological) involving others into cognitive processes involving just the individual (intrapsychological) is the "result of a long series of developmental events. The process being transformed continues to exist and to change as an external form of activity for a long time before definitively turning inward" (Vygotsky, 1978, p. 57). All of these characteristics that contextualize the development of HMFs suggest that the social origin of learning and development is intensively connected to understanding the *zone of proximal development*.

The Role of the Zone of Proximal Development

A critical concept projected by Vygotsky (1978) is the *zone of proximal development*, which is his most referenced construct. It is in the context of this construct in which Vygotsky suggested reconceptualizing intelligence by focusing on intellectual potential that might be difficult to define or measure by a conventional intelligence test. Vygotsky argued for a process-product approach in examining and describing the relation between learning and development. In essence, the zone of proximal development, advocated by Vygotsky, provides "psychologists and educators with a tool through which the internal course of development can be understood" (p. 87).

In particular, the zone of proximal development extends to a comprehensive understanding of development beyond end results, but offers a conduit for examining the very processes that lead to mature functions. It differentiates those "functions that have not yet matured but are in the process of maturation, functions that will mature tomorrow but are currently in an embryonic state. These functions could be termed the 'bud' or 'flower' of development rather than the 'fruits' of development" (p. 86). It is this critical aspect in which the notion of the zone of proximal development is discussed in this section.

The Complex Nature of the Zone of Proximal Development

Vygotsky (1978, 1986) proposed that when an individual participates in joint activities, the social situation transforms the cognitive development of the individual. In order to examine the development of the individual, it becomes necessary to focus on the process, rather than product, of learning. The zone of proximal development provides the milieu for exploring the process of learning and development "in the state of formation." Vygotsky defined the zone of proximal development as *"the distance between the actual developmental level as determined by independent problem solving and the level of potential development as determined through problem solving under adult guidance or in collaboration with more capable peers"* (p. 86). This abstract area is large at the beginning of the task when the learner needs assistance to grasp the new concept. With the assistance of the more knowledgeable other, the zone of proximal development shrinks; the learner needs less and less help with the task, moving eventually to a point of independence. However, this explanation does not illustrate the complex nature of the zone of proximal development. What it fails to emphasize is the intellectual and dynamic nature of the zone of proximal development.

Conducting a large empirical study, Vygotsky investigated the dynamics of intellectual development, concluding that the zone of proximal development is a more inclusive predictor of learners' intellectual development than the conventional IQ score (van der Veer & Valsiner, 1991). In Vygotsky's research and writing, he did not discount the role of the conventional IQ, suggesting that at best, it provided useful information about the learner's independent performance or the learner's actual developmental level. Yet, Vygotsky acknowledged that the conventional IQ poses several problems, which include assessment of the student's potential for learning, the measurement of assisted performance, and the contexts in which learning and development may occur, including collaboration and the role of intersubjectivity.

Learning Potential, Performance, and Context

It is remarkable that Vygotsky formed the concept of zone of proximal development to deal with the issue of using conventional IQ tests; and today we can apply the concept of the zone of proximal development to deal with new forms of quantitative tests that focus solely on a learner's independent performance (Albert, 2002). Studying what learners may accomplish when their performance is assisted by more experienced others provides some sense of their potential. For example, a student's ability to solve a mathematics problem with the help of others (e.g., peers, teacher), which the student would not have been able to solve independently, gives educators insight into the skills and concepts that are in the process of formation, but are not yet fully developed in the individual. In other words, the individual has the potential to develop these skills with proper assistance or scaffolding, which is discussed later in this section.

Thus, it is important for students' educative environment or context to utilize the zone of proximal development. When teachers continually offer students problems they are able to handle without assistance, or provide experiences that are too distant from students' independent level of performance, they are failing to provide instruction that enhances intellectual development. Vygotsky accentuated that teachers must collaborate with students in joint cognitive activities chosen to fit their level of potential development, thereby advancing the student's actual development. Vygotsky argued that as instruction leads to new knowledge and skills, it also permits students to move to new levels of understanding in which they become aware of and take control of their intellectual activities.

To further clarify this idea, let us examine an example from a high school mathematics classroom. An algebra teacher introduces the properties of graphs of exponential functions after the students have learned about the properties of graphs of linear functions. Specifically, the teacher is trying to help the students determine the domain of an exponential function. Rather than working within the students' level of actual (unassisted) performance and asking students to define domain or to name all of the x-values of the function, which they can clearly accomplish independently by reading the table, it is more effective for the teacher to ask a question that is aimed slightly beyond their level of actual development. Thus, the teacher asks students to apply their prior knowledge of the domain of a linear function to solve the problem of finding the domain of an exponential function. By presenting the problem in this manner, the teacher facilitates a useful connection between students' prior knowledge, which they may not have been able to discover on their own. The idea is to help students understand that the domain of a linear function is all real numbers because the graph extends to infinity along the x-axis in both directions, and in the same way, the graph of an exponential function extends indefinitely along the x-axis. Therefore, they would determine that the domain of an exponential graph is the same (i.e. all real numbers). Thus, when the students become aware of this connection on their own, they are able to apply it to novel situations, for example, when they encounter the graphs of quadratic functions. When they are asked to find the domain and range of any function, if they use their prior knowledge without the teacher guiding them to do so, they will have advanced their actual development, taking more control over their own learning.

This example refers to the Vygotskian view that learning leads development, whereby learning and development are neither separate nor identical measures. They are combined in a multifaceted, interconnected manner.

> What [students] can do in cooperation today [they] can do alone tomorrow. Therefore, the only good kind of instruction is that which marches ahead of development and leads it. It must be aimed not so much at the ripe as at the ripening functions. It remains necessary to determine the lowest threshold at which instruction in, say, arithmetic may begin, since a certain minimal ripeness of functions is required. But we must consider the upper threshold as well; instruction must be oriented toward the future, not the past (Vygotsky, 1986, pp. 188–189).

The Vygotskian view that learning leads development implies a very different approach to mathematics classroom instruction than either the *separatist* or the *identity* perspective. Rather than proposing active students who take charge of their own development (*separatist*) or passive students controlled by the surrounding environment (*identity*), the sociocultural historic vision proposes that active students and active socially constructed contexts collaborate to generate developmental change. Development occurs in the zone of proximal development, a phase of mastery created in the course of social interactions in which students have moderately attained skills but can only effectively apply them with the assistance and regulation of an expert collaborator, be it a peer or teacher. The collaborator leads and organizes the activity, *scaffolding* the learner's efforts to a higher level of proficiency or performance. As learners internalize features of this interaction, they work on it and restructure it, striving to comprehend and apply it to similar, but new, situations (Wertsch, 1985, 2008; Wood, 1980).

Wood, Bruner, and Ross (1976) introduced the term *scaffolding* to refer to an essential construct that emerges from the concept of the zone of proximal development and its relationship to teaching and learning. This term is used to describe how performance is assisted in a tutorial relationship among individuals of differing levels of conceptual knowledge. The construct of scaffolding helps us make sense of the role of the zone of proximal development. In particular it answers the question,

> [How] can this child eventually be able to function at the same level independently? Scaffolding answers this question by focusing on the gradual 'release of responsibility' from the expert to the learner, resulting in a child eventually becoming fully responsible for his/her own performance. This gradual release of responsibility is accomplished by continuously decreasing the degree of assistance provided by the teacher without altering the learning task itself. Emphasizing the fact that the learning task remains unchanged makes scaffolding different from other instructional methods that simplify the learner's job by breaking a complex task into several simple ones. (Bodrova & Leong, 2001, pp. 11–12).

Essentially, scaffolding is a tool that supports the learning of specific content (i.e. concepts, skills, or tasks); the scaffold or support is gradually removed until the less experienced learner no longer requires assistance. Once the scaffold is phased out entirely, mastery of the content or task is achieved, allowing the learner to perform the task independently. A much more explicit explanation and application of this construct is presented in Chap. 4 in which we examine teacher learning in a collaborative context.

Intersubjective Learning

Another concept rooted in the zone of proximal development that is essential to understanding Vygotskian theory on the role of collaboration in student learning is *intersubjectivity*. As discussed previously, Vygotsky's zone of proximal development is the difference between what a student can accomplish independently and what he or she can accomplish with the assistance of a more competent person (Tudge, 1992). This theory helps to explain collaborative processes that afford opportunities for learners to work with each other in pursuit of knowledge, skills, and ideas. These collaborative processes create an opportunity through which a group of learners begins a task, activity, or discussion with different understandings but ultimately, through communication, achieve a shared understanding or a "state of intersubjectivity" (Rommetveit, 1979, p. 94). Intersubjectivity results from this interaction as the perspectives of all the learners intertwine, mingle, transform, and coalesce to develop shared meanings. However, recent interpretations suggest that developing "shared meanings" does not necessarily mean that the participants attain "identical conceptual structures," but rather, that "their conceptual structures are sufficiently compatible for successful reciprocal assimilation" (Steffe & Thompson, 2000, p. 193). In other words, communication allows participants to achieve what Rommetveit more appropriately terms "states of partial intersubjectivity" in which their ideas and conceptions are compatible but may not be exactly the same.

To exhibit intersubjectivity and to communicate effectively during joint activity, it is essential that the learners work toward the same goal (Berk & Wensler, 1995; Bruner, 1996). Yet, it is not essential for the parties to remain in a constant state of agreement or to reach a common end solution (Nathan, Eilam, & Kim, 2007). The current understanding of intersubjectivity, referred to as the *participatory view*, emphasizes that both consensual agreement *and* disagreement are important in mediating collective activity. This challenges the traditional view, which strictly judges agreement as favorable and disagreement as unfavorable. In fact, the role of disagreements in cognitive development has been established as a significant one. For example, it was determined in Nathan et al. study that participants' desire to reach a common understanding of a mathematics problem leads them to express their differing ideas in more "refined and accessible ways" (p. 524). Students were presented with the 'Pie Problem', which asked, "How do you cut a pie into eight equal-sized pieces making only three cuts?" (p. 528). They worked on this task first individually, then in pairs, and finally as a whole class. Throughout the discourse, students with divergent points of view challenged one another in respectful and constructive ways, driving each other to present clearer and more articulate arguments. Thus, disagreements fostered critical discourse and led students to develop more sophisticated ways of arguing their differing ideas. The study findings suggest that although no clear convergence in a solution to the problem was achieved, intersubjectivity played a central role in shaping discussion among peers and lead to a clearer understanding. These results support Vygotsky's placement of intersubjectivity at the "heart of learning and consciousness itself" (Nathan et al., p. 524).

The importance of intersubjectivity in social interactions is well documented (Matusov, 1996; Nathan et al., 2007). The participatory view of intersubjectivity focuses on "the coordination of individual participation in joint sociocultural activity" (Matusov, p. 26). Social interactions that occur within the learning community provide the context for shared thinking. It is through social interactions that participants use "communicative tools" to negotiate meaning as they strive for a shared notion of the situation (Albert, 2000, 2002). Intersubjectivity therefore, becomes, "a condition for, or characteristic of, true human communication, implying for the interlocutors a reciprocal faith in a shared experiential world" (Smolka, DeGoes, & Pino, 1995, p. 169).

Achieving intersubjectivity means learners must do more than just work together or allow one person to dominate the activity. They must share power, "where inequality between partners resides only in their respective levels of understanding" (Driscoll, 1994, p. 236). Sharing the power or authority in an interaction reduces the subjective difference between group members. New possibilities and opportunities are opened up for participants, which lead to a better understanding of the activity or discussion. The study of intersubjectivity is becoming increasingly important with the growing awareness of the social nature of human thought and development (Albert & McKee, 2001). Without at least the opportunity for individuals to share and discuss their diverse viewpoints, which may or may not lead to the achievement of intersubjectivity, "we learn nothing, and do little to advance and refine our understanding and our means of communicating our understandings to others" (Nathan et al., 2007, p. 556).

Concept Development in Thinking

Another key domain of Vygotsky's sociocultural historic theory that relates to mathematics teaching and learning is his empirical work on *concept development* (or formation). According to Vygotsky, a concept is a *"complex and true act of thinking* that cannot be mastered through simple memorization" (Vygotsky, 1987b, p. 169). While memorization is often one of the important initial steps in the process of concept formation, many of the processes discussed previously in this chapter, such as the transition from lower to higher mental functioning, mediation through signs and tools, speech and language acquisition, and social or cultural context, also play an essential role during the course of development.

Concept formation is the complex process of generalization and internalization of socially meaningful activities resulting from a dynamic interaction between the concrete and abstract (Harvey & Charnitski, 1998; Smagorinsky, Cook, & Johnson, 2003). In order for a concept to form, a child's mode of thought must be elevated (Vygotsky, 1987a, 1987b). It is the very nature of a concept that makes this process so complex.

> A concept is not just an enriched and internally joined associative group. It represents a qualitatively new phenomenon, which cannot be reduced to more elementary processes, which are characteristic of the early stages of development in the intellect. Concept thinking is a new form of intellectual activity, a new mode of conduct, a new intellectual mechanism. (Van Der Veer & Valsiner, 1991, p. 259)

Concepts are formed as a result of a complex interaction among all the basic intellectual functions. According to Vygotsky, the process of concept development is characterized by two "types of generalizations that approximate concepts yet do not achieve their theoretical unity;" these are called complexes and pseudoconcepts, and they are developed along the course of concept formation (Smagorinsky et al., 2003, p. 1399).

The first step in conceptual development occurs when a child arranges objects or ideas together that are associated with one another but are not logically unified by the same theme or trait. The set of such elements is referred to as a complex. An example of the development of a complex provided by Smagorinsky et al. (2003) is when a child learns to name a canine a dog, but then also labels any other quadruped a dog (p. 1402). The intermediate stage of concept development or bridge between complex and concept is that of the pseudoconcept. Vygotsky referred to a pseudoconcept as a "shadow of a concept," mimicking a concept but slightly different, because the individual elements of the pseudoconcept appear unified but are related based on mere association (p. 1404). Continuing with the previous example, this occurs when a child learns to label a canine a dog, but then also labels a wolf (or anything else dog-like) a dog. Vygotsky provides a useful analogy for the relationship between pseudoconcept and concept, stating that, "the pseudoconcept is as similar to the true concept as the whale is to the fish" (Vygotsky, 1987a, p. 144). Finally, the process ends in the formation of a true concept, which consists of a set of elements unified by one theme. In the course of concept development, the learner follows a multifaceted path, from the complex based on loose association, to the pseudoconcept that appears conceptual yet is inconsistent to the unified concept (Smagorinsky et al., p. 1404). This entire process begins in early childhood and reaches maturity during early adolescence.

An integral part of the process of concept formation is the role of mediating signs, namely the 'word,' as a means by which the children "direct [their] mental operations, control their course, and channel them toward the solution of the problem confronting [them]" (Vygotsky, 1962 as quoted by Meissner, 2008, p. 225). Thus, the word is first used to form the concept and then develops into a symbol (Meissner). In adolescence, the individual reaches an important milestone in the process of concept development, learning to regulate his or her own mental processes and actions using words or signs. Adult speech plays an important role as the child's mental processes develop.

First, the child is only able to act with the adult's assistance (assisted performance), but eventually, through practice and continual reinforcement across various contexts, the child learns how to take control of his own behavior using speech as a support (self-direction). As a concept (word meaning) is internalized, it becomes an "internal function of the child's mind" (Meissner, 2008, p. 227). Thus, the process of concept development proceeds from action to thought. The "central moment" in concept formation is "the point at which the child is able to use words as functional tools," or a means by which to guide mental operations and focus their course toward solving a problem (Harvey & Charnitski, 1998, p. 153). Therefore, once an individual can perform the action or task on his own *and* can describe the method verbally, in written word, or even pictorially, then the process of internalization is complete.

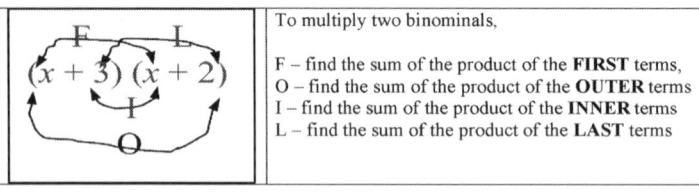

Fig. 2.3 Illustration of the FOIL method for multiplying two binomials

Fig. 2.4 Using algebra tiles
to compute the product of
two binomials

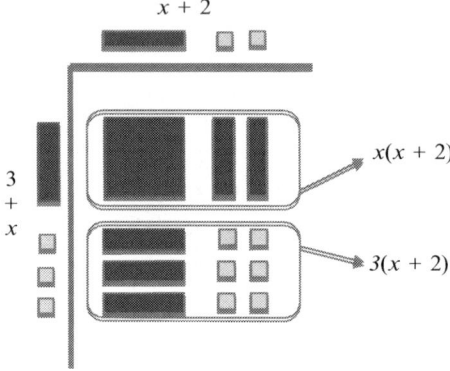

Table 2.2 Example of multiplying two binomials using the FOIL method

Example	Product of first terms	Product of outer terms	Product of inner terms	Product of last terms
$(x+3)(x+2) =$	$(x)(x) +$ $x^2+2x+3x+6$ x^2+5x+6	$(2)(x) +$	$(3)(x) +$	$(3)(2)$

Let us consider the process of internalization in the context of an Algebra I class. A student is learning how to multiply binomials using the *FOIL Method* (See Fig. 2.3). First, the teacher explains the Distributive Property and shows the student how to physically carry out the steps for multiplying $(x+3)(x+2)$ by illustrating the problem with algebra tiles according to Fig. 2.4. Then, she introduces the FOIL Method, explaining what it stands for (first, outside, inside, last) and how it is used as a short cut for the Distributive Property in order to multiply binomials, which is a more direct and mentor model. Table 2.2 shows an illustration of the same problem using the FOIL method. If the student is to begin the process of mathematical concept formation and to develop a basic understanding of multiplying binomials, he or she must practice using the FOIL Method, first under the direction and guidance of the teacher, by working through some examples together on the board (assisted performance). The next step is to practice several different examples independently or with assistance, if necessary, and it is important to vary these examples so the student is forced to think about the steps he or she is performing, instead of merely 'going through the motions' automatically.

For instance, in addition to basic problems like $(x+2)$ $(x-4)$, the teacher is encouraged to give students problems with different coefficients such as $(3x-5)$ $(2x+1)$ and (x) $(5x-3)$, or $(\frac{1}{2}x-2)$ $(\frac{2}{3}x-4)$. These problems require students to use what they know about the Distributive Property to determine how to work with the coefficients when multiplying. At this point, students are still in the 'action stage' of concept formation. They are first able to master the procedural aspect of the Distributive Property, carrying out the operation of multiplication using the FOIL Method. With this mastery comes a lower cognitive grasp of the procedure for multiplying binomials.

Gradually, with enough time, scaffolding, and practice, and with the possible introduction of new methods of conceptualization, (such as the pictorial use of Algebra Tiles as shown in Fig. 2.4) students begin to develop a deeper understanding of *why* the FOIL method works in relation to the Distributive Property. Internalization of this concept is evidenced by a student's ability to provide a concrete (and correct) explanation of each step of the process either verbally, pictorially, or in written word. Thus, conceptual formation is achieved on a higher cognitive level.

Another important indication of conceptual mastery is the ability to generalize what is learned to new situations. For Vygotsky, it is this "transition from one structure of generalization to another" that is at the heart of concept development (Vygotsky, 1987b, p. 170). For example, a student who truly grasps the concept of the Distributive Property will be able to generalize it when faced with a multiplication problem involving polynomials, such as $(4x+9)$ $(3x^2-5x+2)$. She will recognize that the same theory behind the FOIL Method, though not identical, can be applied to this slightly more complicated problem. In other words, she will be able to generalize the Distributive Property to all higher order polynomials that she encounters in the future. Thus, she moves from elementary generalizations (using the Distributive Property only with binomials and with explicit direction) to higher forms of generalization (using this property with all polynomials and variables in the context of any mathematics problem), which is an important step in Vygotsky's process of concept development. The result is the formation of a true concept of the Distributive Property in the mind of the student.

Spontaneous and Scientific Concepts

An important distinction to be made concerning the process of concept development is that of spontaneous (everyday) verses scientific (theoretical) concepts. There are two major differences between the two: (1) Manner of Acquisition, (2) Defining Characteristics (Wells, 1994). The first difference involves the manner in which the concepts are acquired. Spontaneous concepts are those gained informally through social interactions involving family, friends, and others in the individual's environment (Yoshida, 2004). These arise unsystematically and are subjective, as they are influenced largely by the child's social and cultural context. Since everyday concepts arise from specific experiences, they follow an upward developmental path, from that of the concrete towards that of abstraction and generalization (Schmittau, 1993; Vygotsky, 1987a, 1987b). By contrast, scientific concepts are academic concepts that are learned systematically and deliberately through formal education (i.e. classroom instruction). "The development of scientific concepts *begins with the verbal definition,* [which then] descends to the concrete" (Vygotsky, p. 168). Through instruction, the child learns to consciously equate word meanings (concepts) with given verbal expressions in order to make the transition into higher-level thinking (Vygotsky, 1997b; Wells, 1994).

The second distinction between scientific and everyday concepts lies in the core characteristics of each. Scientific concepts have four attributes that are lacking in spontaneous concepts: generality, systematic organization, conscious awareness, and voluntary control (Wells, 1994). The first two are essential to making a concept scientific, because they follow directly from the highly structured classroom environment in which these concepts are developed. Since they are decontextualized, scientific concepts are objective; they hold true in all situations, and thus, are more general and more abstract than everyday concepts. The elements of conscious awareness and voluntary control in scientific thinking develop over time as a "product of the instructional process itself" (Vygotsky, 1987b, p. 169). The systematic nature of the acquisition of scientific concepts, based on the transfer of knowledge from teacher to child in the educational process, is what allows the child to take voluntary control of the concept's formation. This is not the case with spontaneous concepts, because they are gained directly from everyday experiences, and thus, is associated with the specific activities through which they were learned.

Vygotsky identified the greatest distinction between the two concepts as their systematic verses lack of systematic acquisition (Yoshida, 2004). Because everyday concepts are not learned under an organized system, but are instead embedded in a specific context or experience, children may inaccurately use them. Analyzing a series of interviews of second graders in a mathematics classroom, Yoshida makes this distinction. A girl who was asked the meaning of the word *half* responded that it meant to share something equally three ways. Her misconception of the part to whole relation of 1/2 was brought about by her everyday experience of sharing with her two other siblings; an everyday concept that could not be separated from the context in which it was embedded (p. 473). Alternatively, if she were taught in school that to take half of something means to split it into two equal parts, independent of the circumstances of the situation, she would have developed a general, decontextualized understanding of the part to whole significance splitting something in half.

Although scientific and everyday concepts do develop along different paths, this does not mean that they are unrelated. Vygotsky argued that the development of everyday and scientific concepts is not isolated, but rather, each type of concept supports the formation of the other through a shared dependence. While scientific concepts are more sophisticated than everyday concepts due to their rigorous academic nature, they lack the meaningful connection with the child's experience (i.e. concrete contexts) that characterizes everyday concepts (Yoshida, 2004). In the course of development, the two types of concepts must, therefore, undergo an intricate and meaningful integration. Vygotsky (1986) summarizes this important relationship as follows:

> In working its slow way upward, an everyday concept clears the path for a scientific concept in its downward development. It creates a series of structures necessary for the evolution of a concept's more primitive, elementary aspects, which give it body and vitality. Scientific concepts, in turn, supply structures for the upward development of the child's spontaneous concepts towards consciousness and deliberate use (p. 194).

In essence, the systematic nature of scientific concepts clarifies and uplifts the child's understanding of everyday concepts, whereas the concrete and experiential basis of everyday concepts enhances the meaning of scientific concepts (Forman, 2006; Meadows, 2006).

The concept of the passing of time, which is commonly taught in kindergarten, first, and second grade, is an example of how an everyday concept clears the path for scientific concept development. Kindergarteners learn time-related terms such as morning, afternoon, today, yesterday, tomorrow, month, and year informally at home or during other informal activities. Thus, they have a relative understanding of the passing of time in direct relation to their home environment. In the formal classroom setting, kindergarten and first grade students learn how to apply these terms when telling time more precisely and accurately; they also learn to use tools that measure time, such as clocks, calendars, and time lines. The relative understanding of morning, afternoon, and evening that is gained at home helps students to associate the hours precisely and sequentially with each period in the day. For example, a kindergarten teacher may have a student create a timeline of the hours that constitute the morning period of the day (i.e. 6 a.m., 7 a.m., 8 a.m. ... to 11 a.m.) and have the student include pictures that represent activities that he or she does at each of these times (e.g. wake up, eat breakfast, go to school, have morning snack, etc…). The teacher may also show the student what each of these morning hours looks like on a clock. Moving sequentially through the afternoon and evening in the same manner will help the student to develop a more accurate and formal concept of the passing of time.

The same can be done using a calendar to learn the names and sequence of months and the progression of seasons. From everyday experience, the child may understand the temperature change in the seasons by relating the seasons to activities he does in each. For example, he knows that he goes swimming in the summer when it is hot, he starts school in the fall when it is cool, and he plays in the snow in the winter when it is cold, but he may not know exactly which months comprise each season. In school, he formally learns the names and sequential order of the 12 months using a calendar or sequence of pictures. The teacher may have the student match each month with a picture that he drew representing a seasonal activity. By relating daily life experiences and activities to calendar months or clock time (as in the previous example), the everyday concept of the passing of time makes the scientific concept more meaningful to the student, illustrating Vygotsky's position.

In the opposite direction, systematically studying the passing of time in school using a clock or calendar supports the child's ability to recognize the time of day or year for certain daily activities. The child's ability to do so independently enhances structure in the child's life. For example, in understanding the passing of time in a day, the child begins to develop a conscious and deliberate awareness that when the clock reads 6:00 p.m., it is almost time for dinner or almost bedtime when the clock reads 8 p.m., without having to ask his or her parents. To further expand the concept of the passing of time, a teacher may ask students to solve problems involving addition and subtraction of time measurements, such as *how many hours until lunch, if lunch is at 12:00 p.m?* Or *how many days until your birthday?* Repetition and practice across both the home and school contexts integrates scientific and spontaneous concepts like these, paving the way for the 'true concept' to develop its "grounding, coherence, and meaning" (Smagorinsky et al., 2003, p. 1408).

Vygotsky's epistemological notion concerning the influence of pedagogy on students' capacity to acquire knowledge sketches a very fascinating and significant picture of the historical and intellectual foundation of development. It moves us toward a better understanding of concept formation of which pedagogy "is one of the principal sources of the schoolchild's concepts and is also a powerful force in directing their evolution; it determines the fate of [the child's] total mental development" (Vygotsky, 1962, p. 85). The formation of everyday concepts is grounded in knowledge gained through tangible, unswerving, and everyday experiences. In contrast, the formation of scientific or academic concepts emerge from the general to specific in that they are mediated by words through verbal interactions, leading to awareness, organization, and theory, detached from convenient everyday experiences. In summary, sense making or word meaning engages functional knowledge consequential of everyday activities that serves as catalysts for the development of scientific concepts.

The Relationship Between Thinking and Language

The introduction to this chapter and the content that follows illustrates the particular relevance of Vygotsky's sociocultural historic theory, examining how the use of language is fundamental to and interconnected with thought and action. Language mediates our actions as it causes us to order and plan our thoughts (Meadows, 2006). Further, it has been said that "Language, the 'psychological tool ... is perhaps the most potent means of integrating practical ... and symbolic ... knowledge" (Meadows, p. 302). It is not until a child has control of these basic language functions that any further psychological achievements can be made in the realm of higher functions, which was discussed in a prior section of this chapter.

Essential to our understanding of how thoughts function, we must first consider the way in which language develops sequentially. While it is impossible for anyone to prove conclusively what infants think, or if they "think" according to our definition of the word, the seedlings of conceptualization are evident in infants. Some theorists, like Meissner (2008), even question whether or not it is possible to think at all without language. However, many developmental theorists assume that a period of pre-speech in thought development exists, though there are differences in what theorists believe this stage means in terms of consciousness (Freud, 1964; Ogden, 1990; Papaeliou & Trevarthen, 2006; Piaget, 1969; Vygotsky, 1987c, 2004). According to Piagetian theory, this type of thinking is called "egocentric" because while a child is learning the rules of language, he does not fully comprehend the connection between language and communication. Piaget believes that it is not until approximately age 7 that a child begins to completely recognize the function of language as a tool of conversing. From a different school of thought, Vygotsky's theory counters Piaget's conjecture. He uses the construct "private speech" in place of "egocentric speech;" private speech is social in nature, as he views children's first speech as used for communicative purposes. Vygotsky's (1987e) fundamental belief of socialization purports that the child's development is leading away from socialization and towards individuation; thus he does not begin as "egocentric" but as a social being (Meissner, p. 231).

Recent research has tested the Vygotskian concept of the social nature of pre-speech utterances. In a study by Papaeliou and Trevarthen's (2006), a pitch pattern recognition software was created to determine the difference between "communicative" and "investigative" utterances of 10-month-old infants by recording the sounds that they made when their mothers played with them in their homes and when they were left alone. After testing four babies, the results implied that 91.75% of the sounds generated by the babies were for communicative purposes. In conclusion, the researchers assert "These findings confirm that prelinguistic vocalizations might serve both as means of purposeful communication and as a tool of thought. These are the functions later assumed by language" (p. 163).

Once a person has left the early stages of childhood, around the time that the child is seven, Vygotsky believes that private speech becomes internalized; the person's thoughts become introverted and are no longer exclusively for the purpose of communication (Frawley, 1997). This is called "inner speech," defined by Frawley as "social dialogue condens[ing] into a private dialogue for thinking" (p. 95). While this form of thinking is informed by language, it does not mirror language perfectly. Frawley believes that thinking occurs in a more splintered, staccato fashion, rather than through proper grammatical or syntactical means; this language is unique to each person (Van Der Veer & Valsiner, 1991). However, as Frawley observed, it is still closely linked to social speech (Meissner, 2008). Vygotsky illustrates this process himself by claiming, "Thought and word are not cut from one pattern. In a sense, there are more differences than likenesses between them. The structure of speech does not simply mirror the structure of thought" (Vygotsky, 1962, p. 208). Inner speech relates to dialogue because in conversation, the listener must pay attention to what is being said, as the words are first processed on a physiological level. They are then interpreted into the "language" of the listener. However, both processes of internal thinking and dialogical speaking are distinct from written language. Vygotsky even claims that inner speech is completely opposite to written speech (Albert, 2000; Van Der Veer & Valsiner, 1991).

Aside from the processes of speaking and thinking, Vygotsky has unique beliefs about language acquisition, maintaining a cultural slant to his theories. He argues that language contains the generalized contexts, which are the storehouse of cultural experiences (Vygotsky, 2004). Without the aid of language, there would be no basis to communicate the many concepts that we come across in everyday life. For example, if a person is thinking of the color pink by visualizing it, without the aid of language, there would no basis of communicating that concept to another individual. A person can only communicate "pink" to another through linguistic means, by associating it with other known concepts that they share in their environment, like referencing other colors. Simply point to a color without dialogue is senseless and does not explain "pink." However, if someone is wearing a pink shirt and says, "this represents the color pink" in order to establish a frame of reference, the concept of "pink" is reinforced by the linguist association. In spite of this, Vygotsky takes caution to point out that "word without meaning is not a word but an empty sound" (Vygotsky, p. 66), hence highlighting that the link between the word and the object or concept must be clearly drawn for the individual to process. To translate these ideas to a mathematical concept, one may also consider counting to be a similar scenario to the "pink" example. If an individual were to gesture to three objects, that would not necessarily indicate counting until language is attached to it and the individual verbalizes, "One, two, three." The person must hear the language of counting and also cognitively recognize the difference in the sums of one, two, and three in order to form a connection between the words for the numbers and the quantity. Considering these examples, Vygotsky's statement becomes clear to us: "The relationship from thought to word is not a thing but a process, a movement from thought to word and from word to thought ... Thought is not expressed but completed in word" (Vygotsky, 1987d, p. 250). We realize that it is only through "the unity of word and thought" that language informs thinking (Vygotsky).

While attaching meaning to a word is the basis of learning a language, a singular grasp over a word's meaning does not suffice to say that a child understands the full gamet of a word because we know that context alters the meaning. Vygotsky claims, "The child's work on a word is not finished when its meaning is learned" (Vygotsky, 1987e, p. 322). No clearer is the problem of "generalization of meaning" experienced than at the academic level. Consider how the word "difference" varies between the disciplines. In a history class context, this word can indicate a change or departure from a previously established political, social, or economic order. However, when thinking of the word "difference" in a mathematical context, it signifies what remains after a quantity is subtracted from another quantity. It can indicate only a change in amount, rather than a qualitative change or transformation that is usually indicated by a "difference" in a historical sense. Additionally, "generalization" in the language of mathematics is particularly pertinent from an algebraic standpoint, as it is through establishing a formula and conducting experiments and proofs that mathematical theories are created.

Recent work by Tohidian (2009) further explores the theories suggested by Vygotsky decades ago. Called the *Linguistic Relativity Hypothesis*, Tohidian proposes that "language influences the way people perceive and think about the world" (p. 67). Tohidian primary question is whether language influences thought or whether language essentially *is* thought. Ultimately, Tohidian conclusion is dual-pronged: language is the most vital tool in the construction of abstract thought development, and native language directly correlates to every-day, habitual thought. Moreover, Tohidian ends in a decidedly Vygotskian tone, quoting Chaika (1989), claiming, "Language and society are so intertwined that it is impossible to understand one without the other. There is no human society that does not depend on, is not shaped by, and does not itself shape language" (p. 2). Therefore, the modern researcher can draw a straight-line conclusion from the principles laid forth by Vygotsky over six decades ago: language and thought exist in a dynamic relationship, which begins with socialization and ultimately concludes at individuation; it is as crucial to the development of self as it is to the purpose of communication and it opens the door to the creation and progression of society as a whole.

Conclusion

From Vygotsky's work, we have learned that the intellectual development of individuals arises from one's culture, which includes the thought, language, and reasoning processes emerging from social engagement and interactions with others to create a joint knowledge of the culture. Intellectual abilities and processes were studied regarding the historical sequence of events that produced them as they relate to one's culture; therefore, culture is essential to intellectual development. What one thinks about knowledge and the process of thinking itself is acquired from one's culture through the use of tools of intellectual thought. It is contextualized by experiences that help shape concept formation in which the role of the other is essential to construction and co-construction of knowledge. These aspects of sociocultural historic theory are applicable to studying intellectual development in general, especially in children's learning and development, but should not be limited to children alone. It should include understanding learning and development across the life span, and when focusing on teaching and learning, particular emphasis should include the study of mathematical learning and teaching through mediation.

References

Albert, L. R. (2000). Outside in, inside out: Seventh grade students' mathematical thought processes. *Educational Studies in Mathematics, 41*, 109–142.

Albert, L. R. (2002). Bridging the achievement gap in mathematics: Sociocultural historic theory and dynamic cognitive assessment. *Journal of Thought, 37*, 65–82.

Albert, L. R., & McKee, K. (2001). In their own words: Achieving intersubjectivity through complex instruction. In V. Spiridonov, I. Bezmenova, O. Kuoleva, E. Shurukht, & S. Lifanova (Eds.), *The summer psychology conference 2000, the zone of proximal development* (pp. 6–23). Moscow: Institute of Psychology of the Russian State University for the Humanities.

Berk, L. A., & Wensler, A. (1995). *Scaffolding children's learning: Vygotsky and early childhood education*. Washington, DC: National Association for the Educating of Young Children.

Bodrova, E., & Leong, D. J. (2001). *Tools of the mind: The Vygotskian approach to early childhood education* (2nd ed.). Upper Saddle River, NJ: Pearson Education, Inc.

Bodrova, E., & Leong, D. J. (2007). *Tools of the mind: The Vygotskian approach to early childhood education*. Upper Saddle River, NJ: Pearson Education, Inc.

Bruner, J. (1996). *The culture of education*. Cambridge, MA: Harvard University Press.

Chaika, E. (1989). *Language the social mirror*. New York: Newbury House Publishers.

Driscoll, M. P. (1994). *Psychology of learning for instruction*. Boston: Allyn and Bacon.

Forman, E. A. (2006). Engendering a learning motive in East Harlem. Essay review of radical-local teaching and learning. A cultural-historical approach by M. Hedegaard, & S. Chaiklin. *Human Development, 49*, 58–64.

Frawley, W. (1997). *Vygotsky and cognitive science*. Cambridge, MA: Harvard University Press.

Freud, S. (1964). An outline of psychoanalysis. In J. Strachey (Ed.), *The standard edition of the complete psychological works of Sigmund Freud* (Vol. 9, pp. 1–95). London: Hogarth Press. (Original work published 1907.)

Ghassemzadeh, H. (2005). Vygotsky's mediational psychology: A new conceptualization of culture, signification and metaphor. *Language Sciences, 27*, 281–300.

Harvey, F. A., & Charnitski, C. W. (1998). Improving mathematics instruction using technology: A Vygotskian perspective. In *Proceedings of Selected Research and Development Presentations at the National Convention of the Association for Educational Communications and Technology (AECT)*, Louis, Missouri.

Heinich, R., Molenda, M., Russell, J., & Smaldino, S. (1999). Media and instruction (Chapter 1). In *Instructional media and technologies for learning* (6th ed.). Columbus, OH: Merrill.

Holborow, M. (2006). Putting the social back into language. *Studies in Language and Capitalism, 1*, 1–28.

Kondratov, A. (1969). *Sounds and signs*. Moscow: MIR Publishers.

Kozulin, A. (1990). *Vygotsky's psychology: A biography of ideas*. Cambridge, MA: Harvard University Press.

Kozulin, A. (1998). *Psychological tools: A sociocultural approach to education*. Cambridge, MA: Harvard University Press.

Luria, A. R. (1979). *The making of mind: A personal account of soviet psychology*. Cambridge, MA: Harvard University Press.

Matusov, E. (1996). Intersubjectivity without agreement. *Mind, Culture, and Activity, 3*, 25–45.

Meadows, S. (2006). *The child as thinker: The development and acquisition of cognition in childhood* (2nd ed.). New York: Routledge, Taylor & Francis Group.

Meissner, W. W. (2008). The role of language in the development of the self II. *Psychoanalytic Psychology, 25*(2), 220–241.

Nathan, M. J., Eilam, B., & Kim, S. (2007). To disagree, we must also agree: How intersubjectivity structures and perpetuates discourse in a mathematics classroom. *The Journal of the Learning Sciences, 16*(4), 523–563.

Ogden, T. H. (1990). On the structure of experience. In L. B. Boyer & P. L. Giovacchini (Eds.), *Master clinicians on treating the regressed patient* (pp. 69–95). Northvale, NJ: Jason Aronson.

Papaeliou, C. F., & Trevarthen, C. (2006). Prelinguistic pitch patterns expressing 'communication' and 'apprehension'. *Journal of Child Language, 33*, 163–178.

Piaget, J. (1969). *The child's conception of time*. London: Routledge & Kegan Paul.

Ratner, C. (2004). Child psychology: Vygotsky's conception of psychological development. In R. W. Rieber & D. K. Robinson (Eds.), *The essential Vygotsky* (pp. 401–414). New York: Kluwer Academic/Plenum Publishers.

Reiser, R. A. (2001a). A history of instructional design and technology: Part I: A history of instructional media. *Educational Technology Research and Development, 49*(1), 53–64.

Reiser, R. A. (2001b). A history of instructional design and technology: Part II: A history of instructional design. *Educational Technology Research and Development, 49*(2), 57–67.

Rommetveit, R. (1979). On the architecture of intersubjectivity. In R. Rommetveit & R. B. Blakar (Eds.), *Studies of language, thought and verbal communication*. London: Academic.

Schmittau, J. (1993). Vygotskian scientific concepts: Implications for mathematics education. *Focus on Learning Problems in Mathematics, 15*(2–3), 29–39.

Shrock, S. A. (1995). A brief history of instructional development. In G. J. Anglin (Ed.), *Instructional technology: Past, present, and future*. Englewood, CO: Libraries Unlimited, Inc.

Smagorinsky, P., Cook, L. S., & Johnson, T. S. (2003). The twisting path of concept development in learning to teach. *Teachers College Record, 105*(8), 1399–1436.

Smolka, A. B., DeGoes, M. C. R., & Pino, A. (1995). The constitution of subject: A persistent question. In J. Wertsch, P. Del Rio, & A. Alvarez (Eds.), *Sociocultural studies of mind* (pp. 165–184). New York: Cambridge University Press.

Steffe, L. P., & Thompson, P. W. (2000). Interaction or intersubjective? A reply to Lerman. *Journal for Research in Mathematics Education, 31*, 191–209.

Stetsenko, A. (2004). Scientific legacy tool and sign in the development of the child. In R. W. Rieber & D. K. Robinson (Eds.), *The essential Vygotsky* (pp. 501–537). New York: Kluwer Academic/Plenum Publishers.

Subbotsky, E. (1996). Vygotsky's distinction between lower and higher mental functions and recent studies on infant cognitive development. *Journal of Russian and East European Psychology, 34*(2), 61–66.

Tohidian, I. (2009). Examining linguistic relativity hypothesis as one of the main views on the relationship between language and thought. *Journal of Psycholinguistic Research, 38*, 65–74.

Tudge, J. R. H. (1992). Processes and consequences of peer collaboration: A Vygotskian analysis. *Child Development, 63*, 1364–1379.

Van der Veer, R., & Valsiner, J. (1991). *Understanding Vygotsky: A quest for synthesis*. Cambridge, MA: Blackwell Publishers.

Vygotsky, L. S. (1960). The development of higher mental functions [Quoted in J. V. Wertsch (1985) *Vygotsky and the social formation of mine*]. Cambridge, MA: Harvard Press.

Vygotsky, L. S. (1962). *Thought and language*. Cambridge, MA: The MIT Press.

Vygotsky, L. S. (1978). *Mind in society: The development of higher psychological processes*. Cambridge, MA: Harvard University.

Vygotsky, L. S. (1986). The development of scientific concepts in childhood: The design of a working hypothesis. In A. Kozulin (Ed.), *Thought and language* (pp. 146–209). Cambridge, MA: The MIT Press.

Vygotsky, L. S. (1987a). An experimental study of concept development. In R. W. Rieber & A. S. Carton (Eds.), *Problems of general psychology: Vol. 1. Collected works of L. S. Vygotsky* (pp. 121–166). New York: Plenum. (Original work published in 1934.)

Vygotsky, L. S. (1987b). The development of scientific concepts in childhood. In R. W. Rieber & A. S. Carton (Eds.), *Problems of general psychology: Vol. 1. Collected works of L. S. Vygotsky* (pp. 167–241). New York: Plenum. (Original work published in 1934.)

Vygotsky, L. S. (1987c). The genetic roots of thinking and speech. In R. W. Rieber & A. S. Carton (Eds.), *Problems of general psychology: Vol. 1. Collected works of L. S. Vygotsky* (pp. 101–120). New York: Plenum.

Vygotsky, L. S. (1987d). Thought and word. In R. W. Rieber & A. S. Carton (Eds.), *Problems of general psychology: Vol. 1. Collected works of L. S. Vygotsky* (pp. 243–285). New York: Plenum.

Vygotsky, L. S. (1987e). Lecture 3: Thinking and its development in childhood. In R. W. Rieber & A. S. Carton (Eds.), *Problems of general psychology: Vol. 1. Collected works of L. S. Vygotsky* (pp. 311–324). New York: Plenum.

Vygotsky, L. S. (1997a). Analysis of higher mental functions. In R. W. Rieber (Ed.), *The history of the development of the higher mental functions: Vol. 4. Collected works of L. S. Vygotsky* (pp. 65–82). New York: Plenum. (Original work published 1931.)

Vygotsky, L. S. (1997b). Consciousness as a problem for the psychology of behavior. In R. W. Rieber & J. Wollock (Eds.), *Problems of the theory and history of psychology: Vol. 3. Collected works of L. S. Vygotsky* (pp. 63–79). New York: Plenum. (Original work published 1925.)

Vygotsky, L. S. (1997c). *Educational psychology*. Boca Raton, FL: CRC Press. (Original work published 1926.)

Vygotsky, L. S. (1998). The development of scientific concepts in childhood. In R. W. Rieber (Ed.), *Collect works: Vol. 5. Child psychology* (pp. 167–241). New York: Kluwer/Plenum.

Vygotsky, L. S. (2004). Thought and word. In R. W. Rieber & D. K. Robinson (Eds.), *The essential Vygotsky* (pp. 65–110). Boston: Kluwer Academic/Plenum Publishers.

Vygotsky, L. S., & Lauria, A. R. (1994). *Studies on the history of behavior: Ape, primitive, and child* (V. I. Golod & J. E. Knox, Ed. & Trans.). Hillsdale, NJ: Lawrence Erlbaum Associates. (Originally published in Russian 1930)

Wells, G. (1994). Learning and teaching "scientific concepts": Vygotsky's ideas revisited. In *Proceedings of the "Vygotsky and the Human Sciences" Conference,* Moscow.

Wertsch, J. V. (1985). *Vygotsky and the social formation of mind.* Cambridge, MA: Harvard University Press.

Wertsch, J. V. (2008). From social interaction to higher psychological processes. *Human Development, 51*(1), 66–79. Retrieved from http://www.karger.com/doi/10.1159/000112532

Wood, D. (1980). Teaching the young child: Some relationships between social interaction, language and thought. In D. Olson (Ed.), *The social foundations of language and thought* (pp. 281–296). New York: Norton.

Wood, D., Bruner, J. S., & Ross, G. (1976). The role of tutoring in problem solving. *Journal of Child Psychology and Psychiatry, 17*, 89–100.

Yoshida, K. (2004). Understanding how the concept of fractions develops: A Vygotskian perspective. *Proceedings of the 28th Conference of the International Group for the Psychology of Mathematics Education, 4*, 473–480.

Chapter 3
Images and Drawings: A Study of Prospective Teachers' Perceptions of Teaching and Learning Mathematics

Introduction

The *Common Core State Standards for Mathematics Initiative* (CCSS) proposes that the teaching of mathematical content needs to underscore both procedural skills and conceptual understanding "to make sure students are learning and absorbing the critical information they need to succeed at higher levels" (CCSS, 2010). Prospective teachers may need to develop not just a deeper knowledge of subject matter (algebra, geometry) but an understanding of the mathematical process of inquiry and problem solving to enrich their teaching practices and to encourage critical thinking skill development in their students. Prospective teachers' perceptions of what constitutes good mathematics instruction pose great influence on the type of mathematics instruction they will deliver in their own classrooms (Hill, 2004). It is not always easy to help prospective teachers gain a concrete understanding of abstract mathematical concepts and operations. Their prior experiences in learning and teaching mathematics may have been based on a symbolic rather than concrete representation of numbers, founded on the rote memorization of rules or formulas rather than on the problem solving process. Rote learning may have assisted teachers and learners of mathematics in performing the operations of addition, subtraction, multiplication, and division correctly; but in learning to give the "correct" answer, they may have never understood the critical thinking process of problem solving in mathematics.

An earlier version of this chapter titled *Prospective Teachers' Perception of Teaching and Learning Mathematics Through Images and Drawings* was presented at the 2005 annual meeting of the American Chapter of the International Group for the Psychology of Mathematics Education, Roanoke, Virginia and was published in the conference proceedings.

In the pursuit of improving mathematics teaching, three questions emerged:

1. How do educators provide opportunities for prospective teachers to gain insight into their individual mathematical learning experiences and teaching practices?
2. Are prospective teachers' images of teaching mathematics embedded within a larger sociocultural framework that extends across time, people, places and specific mathematics content?
3. Did the prospective teachers use newly acquired knowledge about mathematics pedagogy to construct ideas of future teaching-learning episodes?

The study presented in this chapter uses drawings and narratives to serve as "communicative tools" for interrogating prospective teachers' learning histories and thoughts about the mathematics teaching-learning process to answer these questions. Furthermore, the drawings present a visual perspective and design that are interconnected in a specific way to the written narratives rather than to independent variables as in traditional research (Scott, 1994).

Drawings and narratives may uncover perceptions of prospective teachers' prior personal teaching/learning experiences in mathematics and provide rich material for self-reflection and analysis of the effectiveness of their teaching strategies. We purport that prospective teachers' drawings and narratives about mathematics generated during their academic experience may play a role in their preparation and development as effective mathematics teachers. This study examines prospective teachers' perceptions of teaching and learning mathematics through a series self-made drawings and narratives that asked teachers to illustrate mathematics teaching situations of their past, present, and idealized future classrooms. These drawings revealed teacher perceptions of old, new, and idealized mathematical teaching experiences, as well as an evolving awareness of pedagogy and curricula that constitute effective mathematics instruction.

In this chapter, we argue that drawings can serve as a substantial tool in raising the quality of mathematics teacher preparation. During their academic coursework and practica program, the prospective teachers are developing a conception of themselves as mathematics teachers. Drawings and narratives can be used to assist the questioning of tacit assumptions that underlie personal pedagogical practices, allowing prospective teachers to improve those practices as they progress in their understanding of real-world aspects of mathematics teaching. Manning and Payne (1993) suggest the development of higher cognitive processes in teachers occurs "not simply [in] quantitative increments but [also through] qualitative shifts as the unique past experiences and previous knowledge of individuals interact with the present learning event" (p. 362). The authors propose the development of a teacher through a learning theory supported by Vygotsky's sociohistorical perspective on knowledge acquisition. We concur that such a theory enables teachers to engage in self-reflection and self-regulation based on a personal sociocultural perspective of their individual learning experiences (Albert, 2000). Drawings and narratives may serve as a catalyst for self-reflection that would be the basis for the emergence of qualitative changes in individual teaching practices in mathematics.

We begin the discussion with a brief summary of relevant research literature on student-generated drawings as well as teacher-generated drawings. Next, we describe the methods and procedures for data collection, which includes a discussion of the qualitative and quantitative techniques employed in the analysis of the drawings and narratives. The final section focuses the importance of the findings to research and practice, including a discussion of these findings.

An Emerging Framework

The research on the use of images and drawings suggests that drawings and images have become more pervasive in daily life (Dickmeyer, 1989; Kendrick & McKay, 2001; Richards, 2006). They reflect the existence of certain values both in our culture and in the standards and norms of our educational system, oftentimes unmasking the conflicting realities that coalesce with today's shifting culture. Kendrick and McKay assert that today's culture is shifting from one dominated by language "to one in which images are becoming increasingly important" (p. 125). It follows from the prevalence of visual imagery that today's students are more visual than those of any previous generation, because of the growing presence and use of multi-media technologies available. Because classrooms exist as microcosms of society, the teaching and learning that occur therein should reflect this emerging importance society places on images. In fact, from elementary school to the college classroom, students have depicted images in drawings that provide insight into their perceptions of teaching and learning processes (Black, 1991; Goodenough, 1926; Gulek, 1999; Haney & Gulek, 1996; Mason, Kahle, & Gardener, 1991; Weber & Mitchell, 1995, 1996; Wheelock, Bebell, & Haney, 2000).

In the 1990s, numerous researchers studied the use of metaphors and images in education (e.g., Bullough, Knowles, & Crow, 1991; Dickmeyer, 1989; Elbaz, 1991; Haney, Russell, Cengiz, & Fierros, 1998; Miller & Fredericks, 1988; Weber & Mitchell, 1995, 1996). This large body of research indicates that the use of drawings enhance the teaching and learning processes in two specific ways. First, student-generated drawings allow the drawers to make their conceptual understanding concrete. In many cases, "these drawings not only show the students' understanding of [learning events and content] but also an understanding of the larger issues represented by these events" (Hibbing & Rankin-Erickson, 2003, p. 762). Second – and most useful to educators – student-generated drawings provide teachers with a tool that allows them to visually assess student understanding. In both situations, drawing proves to be a useful tool in the teaching and learning processes.

Prior inquiries into student-generated drawings set the stage for future researchers. Despite success in using student-generated drawings to study student perceptions at the K-12 level, little research has been done to examine the effectiveness of using drawings to study teacher learning. Haney, Russell, and Bebell (2004) call drawings a "seriously underdeveloped line of inquiry about teaching and learning" (p. 266). Haney et al. assert that drawings level the field for both educator and learner regarding "adequacy of expression" (p. 241) providing an avenue for prospective teachers to freely express their perceptions and understandings of pedagogy and instructional practice, as well as a tool for instructors to gauge these understandings and perceptions. When analyzing drawings, the focus is not on the artistic or aesthetic nature of the drawings, but rather on how the drawers perceive, organize, express, and give meaning to an emerging conception of themselves as learners and teachers of mathematics. Not only does this analytical focus allow researchers to use drawings as a tool to uncover prospective teachers' conceptions of mathematical learning, but it also allows instructors to harness "the power of assessment to effect change and improvement" (p. 246). Teacher-generated drawings and narratives can serve as internal "communicative tools" for mediating inner thoughts about mathematics teaching and learning, as well as tools that facilitate external communications, with others (Albert & Rhodes, 2005). Our goal is to use drawings as tools to comprehend how prospective teachers make sense of their teaching over time as they experience learning mathematics in different sociocultural contexts, e.g., courses and practica.

Messages in drawings are more figurative than literal. Drawings do not always represent hard-nosed reality of teaching and learning situations. Content in drawings may be the result of preconceived notions, stereotypes, or dramatic one-time occurrences (Haney et al., 1998; Olson, 1995). The possibility exists that our prospective teachers drew what they perceived to be expected of them. Haney (in Haney & Gulek, 1996) recommends that large random samples of drawings be used for the purpose of analyzing patterns and trends (Olson). Although drawings may not reveal a precise view of reality, collections of drawings can provide a unique look at classrooms from the viewpoints of the participants – in this case, prospective teachers. "Research says kids' drawings are affected by preconceptions but also by experience. In a lot of feedback sessions, teachers often start by saying, 'This is the result of the children's preconceptions,' but then they see drawings that reflect real experiences unique to their school," (Haney as quoted in Tovey, 1996). Bolstering this analysis are written descriptions of the drawings provided by each of the prospective teachers as well as the transcripts of focus groups interviews conducted with the prospective teachers while they were presenting their drawings to their peers. As in Wheelock, Bebell, and Haney's (2000) study, written and oral information were used to corroborate the findings from the drawing analysis.

The epistemological approach of Vygotsky's (1978) is useful in understanding prospective teachers' drawings. Vygotsky describes the role that culturally-bound "tools and signs" (pp. 52–55) have in assisting the development of the human mind in its learning and thinking, theorizing that such tools and signs are a deep reflection of psychological processes, directing us to a deeper understanding of the activity in which we engage. The zone of proximal development permits growth of independent intellectual functioning through the "actual verbal interaction with a more experienced member of society via the richness and substantiveness of verbal dialogue" (Manning & Payne, 1993 p. 364). Vygotskian theory presents a powerful conception of human learning represented in drawings whereby the sociocultural context exerts a strong influence over these teacher-generated drawings and narratives. Drawings accompanied by narratives communicate the subtleties of emerging

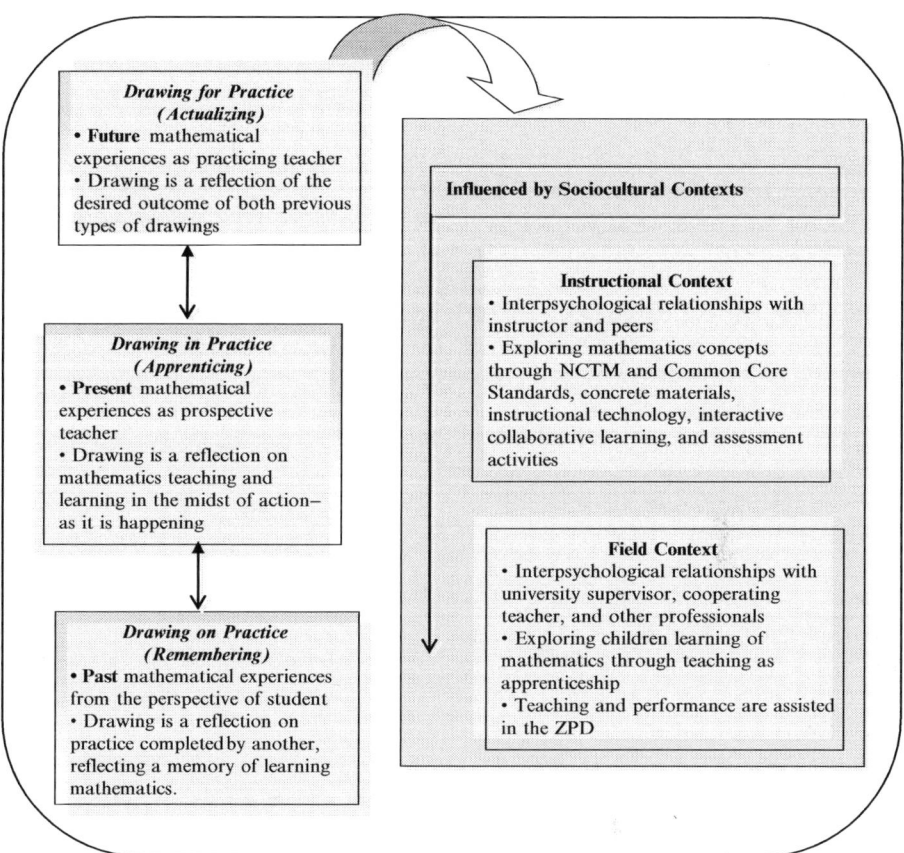

Fig. 3.1 Theoretical framework for data collection and analysis

understandings of prospective teachers' conceptions of themselves as mathematics teachers, helping them clarify, organize, and come to terms with pedagogical knowledge and an image of themselves as mathematics teachers. Thus, the major purpose to analyze teacher drawings and narratives "is to resolve the unexpected, to settle the [teachers'] doubt, or in some manner to redress or explicate the imbalance" (Bruner, 1996, p.121) prompted by the drawings.

Figure 3.1 shows the theoretical framework that guided data collection and analysis of the drawings, narratives, and focus groups interviews. This framework includes three unique experiences: past reflection of mathematical experiences as a student (*Drawing on Practice, Remembering*), current experience as a prospective teacher (*Drawing in Practice, Apprenticing*), and the future image of the student in the role of the practicing teacher (*Drawing for Practice, Actualizing*). The primary influences on these practices are the instructional and field contexts. The framework depicts the transforming agents of prospective teachers' conceptions of mathematics teaching from past-to-present-to-future, describing collaborative influences in which the individual and the sociocultural contexts are considered.

Thus the first two in the series of three drawings and accompanying narratives analyzed in this study provide a vista into the past and present experiences of the prospective teachers; the final drawing, *Drawing for Practice (Actualizing)*, becomes an important connection to the course and practica. It provides insight on whether and to what extent the prospective teacher's learning is influenced by mental representations and material acquired in their teaching courses. These drawings frame the learning that the development of new knowledge and experiences create. The frames serve as tools of reflections and examples of the practical knowledge gained by the learners. They are more than just words; they are images that have theoretical, pedagogical, and communicative components.

Methodology

This study of 240 drawings was conducted in an undergraduate teacher education semester-long elementary mathematics pedagogy course; all 80 prospective students were enrolled in three sections of the course. During the course, the prospective teachers were asked to draw a series of three pictures in response to three different prompts. The prompts were given to all of the prospective teachers in the course at once and were delivered in a set order. The first prompt was given at the start of the course, the second towards the middle of the course, and the third at the end of the course. Each prompt included instruction to provide a brief narrative of mathematics teaching and learning portrayed in the drawing. The prompts, in order, were:

1. Think back to when you were a third-grader in mathematics class. Think about the things you did in that class, what it was like to be a student in that mathematics class. When you have an image of that, draw a picture of yourself in your third grade mathematics class.
2. Think about the things that are currently going on in your prepracticum mathematics classroom. Think about the mathematics that is being taught and the instructional methods that you're witnessing. Draw a picture of what is going on in that mathematics classroom.
3. If I were to visit your classroom 5 years from now, what would mathematics teaching and learning look like in your classroom? Draw a picture of what a visitor would see in your classroom.

These prompts evoke different time periods and different stages in the prospective teachers' lives. The first asks them to conjure an image from the past wherein they were the ones receiving the instruction. These pictures depict *memories* of prior curricula and instruction. The second prompt asks prospective teachers to depict their current circumstance as student teachers in a mathematics practicum.

The immediacy of these drawings summons images of current teaching *experiences as apprentices*, where reality was more likely depicted than in the other drawings. The third prompt asks the prospective teachers to imagine their classrooms in the future – in which they are completely in charge. This prompt likely evoked images of *actualized* teaching, in addition to demonstrating the growth in pedagogical and instructional practice acquired in the elementary mathematics methods course.

The timing of the prompts interacts with prospective teachers' learning in their mathematics course, a course in which they experiment with various pedagogical and curricular forms that influence their perceptions of deliberate practice in mathematics. An essential element of the course is its field-based component. First, all students enrolled in the course complete a prepracticum at an urban school. Second, several of the assignments include a field-based aspect. For example, students design and put into practice lessons and activities that require the use of manipulatives, problem solving, and assessment techniques explored in class. They then write a critical reflective narrative about the lesson content and its affects on student learning. Because of this course focus on teaching and learning mathematics content, an evolution in prospective teachers' perceptions is expected to be apparent across the drawings.

To help understand and interpret the drawings and accompanying narratives, 14-member focus group from each of the three course sections examined and commented on a representative sample of drawings and narratives as well as provided their interpretation; appropriately 30 drawings, completed by ten participants, represented the three prompts. The focus group interviews were designed to elicit information about the prospective teachers' experience in this inquiry, to explore their "meaning-making" processes, and to differentiate the "details of their experience from their stream of consciousness" (Seidman, 1991, p. 1). This process involved the focus group's participants searching for patterns across the drawings and narratives, describing what the drawing and narratives meant, and reflecting on how these meanings may lead toward a deeper understanding of teaching and learning mathematics. The focus group interviews were taped-recorded and transcribed verbatim.

Coding Drawings

Procedures used in this analysis are outlined in Haney et al. (1998). First, two coders viewed the drawings separately to create a coding scheme. Next, coders worked to develop one set of code definitions that captured the information in the separate lists. These codes and definitions are provided in Appendix A. The researchers then coded a subset of drawings separately to test the consistency of the coding scheme. Inter-rater reliability estimates were then computed for each code section and the coding scheme was revised to improve on these estimates. When the checklist was finalized, inter-rater reliability estimates were computed on a random sample of 20 drawings. Kappa estimates ranged from a low of .73 on several of the subsections to

Table 3.1 Inter-rater reliability and coefficients

Rating scale section	Inter-rater reliability coefficient	Cohen's Kappa
Teacher	0.93	0.74
Teacher affect	0.80	0.74
Students	0.94	0.75
Student affect	0.95	0.93
Desks	0.81	0.73
Furnishings	0.96	0.79
Mathematics concepts	0.98	0.86
Tasks/activities	0.87	0.73
Level of representation	0.89	0.73
Manipulatives	0.96	0.75

a high of .93 on "student affect," indicating good inter-rater reliability (Kvalseth, 1991; Uebersax, 2002). Coefficients of simple inter-rater reliability ranged from a low of .81 for "desks" and a high of .96 for "manipulatives." Kappa estimates ranged from a low of .73 on several of the subsections to a high of .93 on "student affect."

Since Kappa almost always yields lower coefficients than a simple measure of inter-rater reliability, researchers for interpreting Kappa have offered several rules of thumb. First, Kvalseth (1991) suggests that a Kappa of above .60 indicates good reliability. Second, Uebersax (2002) cautions that Kappa may be misleadingly low and is difficult to interpret. He suggests that a negative Kappa indicates agreement less than chance, while a positive Kappa indicates agreement that is greater than chance. The ranges for this study indicate good inter-rater reliability. Table 3.1 contains all of the reliability coefficients calculated for the study, rounded to the nearest hundredth. All of these are higher than Kvalseth's suggested criterion of 0.60, indicating good inter-rater consistency. Finally, the codes were entered into an Excel database, and frequencies and percentages were tabulated for each of the codes, as well as for the composite variable, as shown in Table 3.1. Preliminary results were checked against the focus group interview transcripts to see if they held up.

Analysis of Focus Group Transcripts

To avoid errors such as the researchers' predisposition about the inquiry, a systemic approach was applied to the analysis of focus group data. This approach utilized multiple readings of transcripts and notes as well as listening to tapes and examining transcripts for patterns among participants' responses, then developing codes and sorting them into categories (e.g., changing roles of the teacher, mathematics representations). Throughout this process, participants' perception of events and experiences were measured to provide a direction for understanding and validity regarding "talk, text, [drawings], interaction, and interpretation" (Riessman, 1993, p. 8). To provide for validity, Krueger (1994) contends that the focus group analysis procedure requires careful consideration of words, tone, and context of participants' responses. In this study, Kruger and Riesman's notion of validity regarding establishing "trustworthiness" was applied to determine whether the prospective teachers recognized the interpretations of the drawings as appropriate representations of their experiences. Thus, the importance of this approach was that it placed emphasis on "openness to a variety of meanings [and the sociocultural] context in which they were created" (Malchiodi, 1998, p. 35).

Results

Findings are reported by Phase (drawings analyses from the first, second, or third prompts), and by the following categories of response: the changing roles of teachers; student affect and pedagogy; trends in representation and content; and changes in room structure.

The Changing Roles of Teachers

Teachers' roles changed dramatically from Phase 1 to Phase 3. Drawings in Phase 1 generally showed teachers at the chalkboard or in the front of the room (66%), with most teachers instructing the whole class (55%). By Phase 3, drawings depicted teachers instructing groups and individuals more often than whole classes (29% vs. 21%); only 18% were depicted at the chalkboard or in front of the class; and 45% of the teachers shown were moving in the classroom or walking toward students (up from 8% in Phase 1). Here we see the changed perceptions of teaching among these prospective teachers – from one of teacher as knowledge transmitter to teacher as facilitator or scaffolder. Prospective teachers' drawings and comments allow us to understand the importance of this change in perception as shown in Fig. 3.2. In Fig. 3.2 (Phase 3 Drawing), one can see the teacher in the middle of the room, apparently working with a group of students.

The classroom layout, as well as its mission is geared toward students learning together. The words "collaboration" and "scaffolding," commonly discussed in this mathematics methods course, hold great significance in this student's depiction of an ideal classroom. For scaffolding to occur, students must enter what Vygotsky (1978) referred to as the zone of proximal development, in which they are just beyond their independent capabilities. Once the student is in this zone, a more knowledgeable learner can support the student to move to the next level of learning. This is one of the main reasons that the positioning of the prospective teacher amidst a group of students emphasizes the use of scaffolding, as the teacher works with students in the zone of proximal development, which is a primary reason why the drawings' progression of teacher positions from chalkboard or desk to moving with students is such an important development.

In my third grade classroom, there was only individual math work. When I think back to my third grade math lessons, I remember completing workbook assignments while sitting at my desk. In this scene, I drew such an assignment. The students are sitting at their desks silently completing the workbook pages. The teacher sits at her desk and grades papers. The students can ask the teacher questions but she never walks around the room to check on their work.

[Phase 1]

In five years, I plan to be teaching in my classroom. My class will be organized in islands of student desks with learning centers and reading tables. I drew my students working in groups on the floor. They are using manipulatives because [these] will be a big part of my program. My attitudes [toward] math has changed a little. I now better understand the importance of manipulatives in math. I also understand the significance of collaboration and scaffolding in math.

[Phase 3]

Fig. 3.2 Phase 1 and phase 3 drawings (changing role of the teacher)

In an effort to capture the changing roles of teachers, composite variables were created to capture both "Traditional Instruction" and "Collaborative Instruction" pedagogies. The Traditional Instruction pedagogies reinforce student passivity, abstract learning and student memorization, and teacher-led whole group instruction, as denoted through the following codes: Teacher at chalkboard/in front of room, Teacher instructing whole class, Teacher at desk, Students sitting at desks/tables, Individual seatwork, Worksheets, Flashcards, Competition, Clock/time, and Abstract representations.

Representations in mathematics instruction are ways to display mathematical concepts. For example, one could show multiplication through just its numerical form: $5 \cdot 5 = 25$. This is an example of an abstract representation as the learner is provided with numbers only, and must associate a meaningful mental image with the numbers to comprehend fully the number sentence. By contrast, concrete representations are those that connect physical objects with the mathematical concept. An example of this is using two dimensional tiles to form a square that has, as its dimensions, $5 \cdot 5$, as illustrated with square tiles in Fig. 3.3. Concrete models allow students to understand mathematical concepts through actual physical experiences, rather than through vicarious experiences. Pictorial representations are two-dimensional depictions of concrete representations.

Fig. 3.3 Five-by-five-
concrete representation with
colored tiles

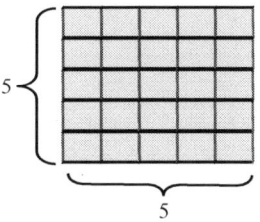

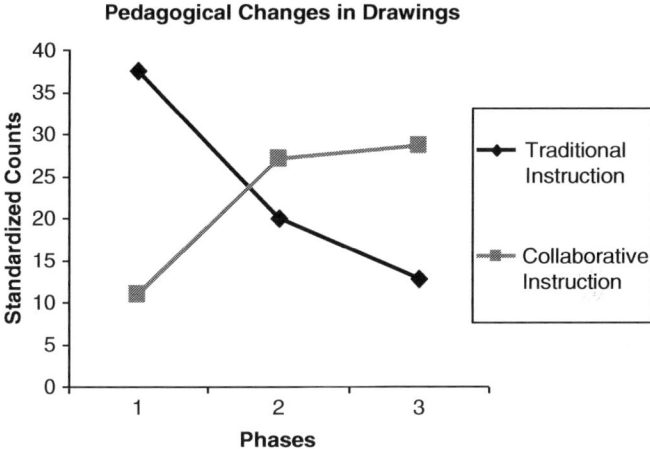

Fig. 3.4 Traditional versus collaborative instruction (pedagogy)

Another composite variable that encapsulates the changing roles of teachers is *Collaborative Instruction*. The Collaborative Instruction composite variable was created to capture pedagogies that support social-constructivist learning. Social learning contexts are those in which students are encouraged to work collaboratively toward a learning goal. Constructivist methodologies allow students to form their own understandings of what is learned. Both of these methods emphasize active student participation and teacher facilitation. Therefore, the Collaborative Instruction composite variable contains the sum of the following codes: Teacher in the center of the classroom, Teacher instructing a group/individual, Teacher sitting in a chair instructing students, Teacher walking around the room, Students at centers, Students talking to each other, Students walking around the room, Active learning tasks, Cooperative learning, Interdisciplinary learning, Tasks with manipulatives, Two-plus activities at once, Concrete representations, and Two-plus representations.

Code frequencies across the Traditional Instruction and Collaborative Instruction variables were made comparable (creating a standard count) by dividing the sum in each Phase by the number of drawings in that Phase and then by the number of codes in that Phase. A similar method of standardizing counts was used for the other composite variables shown in Fig. 3.4, which depicts a consistent drop in Traditional Instruction pedagogies in each Phase. Similarly, Collaborative Instructional pedagogies

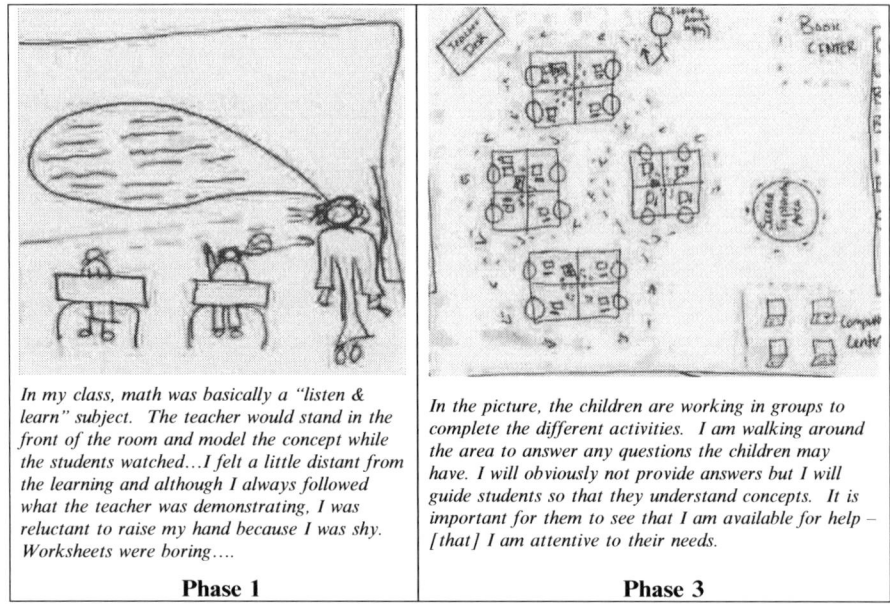

In my class, math was basically a "listen & learn" subject. The teacher would stand in the front of the room and model the concept while the students watched...I felt a little distant from the learning and although I always followed what the teacher was demonstrating, I was reluctant to raise my hand because I was shy. Worksheets were boring....

Phase 1

In the picture, the children are working in groups to complete the different activities. I am walking around the area to answer any questions the children may have. I will obviously not provide answers but I will guide students so that they understand concepts. It is important for them to see that I am available for help – [that] I am attentive to their needs.

Phase 3

Fig. 3.5 Phases 1 and 3 drawings with accompanying narrative (two teaching methods)

depicted in the drawings show a sharp increase from Phase 1 to Phase 2 (and a smaller increase from Phase 2 to 3). This trend evidences one of the following two scenarios (or a bit of both): (a) prospective teachers have become more cognizant of Collaborative Instruction pedagogies through their course work, and/or (b) mathematics classrooms invoke Collaborative Instructional practices more today than in the past. The sharp increase in Collaborative Instruction codes from Phrase 1 to Phase 2 suggests the second scenario.

To better illustrate the two teaching methods, drawings exemplifying each instructional method are shown in Fig. 3.5, along with excerpts from the accompanying texts. The contrasts between the drawings and excerpts point to crucial differences in student affect between the two approaches. Here, prospective teachers associate traditional methodologies with student isolation. In each, students tend to be on their own, adrift in the classroom.

Learning and engagement are associated with aptitude and inner drive in these classrooms, not with pedagogy. As one prospective teacher put it:

> … After you passed that [multiplication] quiz you would get a sticker in that number's column on the math poster, so everyone would know if you knew your times tables. When everyone had passed, we would have a pizza party. This was extremely frustrating and sad to me because at the end of the unit it was only me and another boy who had not learned [the] times tables. Because of us, we did not have a party. At 3rd grade I already knew I was a failure at math. It made me feel stupid.

Collaborative drawings depicted much more inclusion, regardless of ability levels. Prospective teachers cited pedagogical practices – cooperative learning, the use of manipulatives, teacher facilitation – as the force for the inclusion of these students. One prospective teacher wrote:

> ... I liked this lesson a lot because I thought it showed some essential qualities of good math lessons. First, it required students to collaborate causing them to share their ideas and to learn from one another. Second, the manipulatives made the lesson interactive. The cubes engaged the children. Third, the cubes aided in the children's transitions from concrete to abstract mathematical understanding. Fourth, the lesson was fun and the students enjoyed it, making the lesson more memorable and effective.

When asked to reflect on the changes seen between the early and later drawings, one prospective teacher noted in an interview,

> These were the first ones and this is a reflection of our own experiences of when we went to school and then it was very much teacher-centered, not student-centered. It was very much about working individually, repetition, drilling; that's what I remember. I think [teaching] has changed a lot. The teacher being in the front of the room is the most [compelling feature of these].

Student Affect and Pedagogy

The relationships between teaching pedagogy and student affect observed the previous section is borne out through a number of significant relationships among student affect and instructional pedagogy are observed in Table 3.2. Basically, a positive relationship exists between Traditional Instruction and negative student expressions based on facial expressions in the drawings or expressed in the narratives ($r = .331$, $p < .01$), while a negative relationship exists between Collaborative and positive student expressions ($r = .229$, $p < .05$).

Interviewed perspective teachers noted differences in student affect between the Phase 1 and Phase 2 drawings. The following comment captures the essence of the perspective teachers' view:

> ... and children are happy in these drawings, too. That's another thing. In the old ones (Phase One drawings), there's no [facial] expression on the students. And look at how detailed these are. They're showing movement and diversity. So that is something that really jumped out at me. The type of math that was taking place, which they're looking at interdisciplinary approaches, that there are computers in the classroom, that there are centers where students are more in the centers. So, it's a much more informal, relaxed atmosphere.

Figure 3.6 shows that negative student feelings did, indeed, decrease over time, from 16% at Phase 1% to 0% at Phase 3. Positive feelings increased sharply from Phase 1 to Phase 2, falling back at Phase 3, largely because student facial expressions were not depicted as often in the last set of drawings.

The interviewed perspective teachers noted a further observation regarding student affect. While students appeared as interchangeable objects in the Phase 1 drawings, many possessed distinctive features in the Phase 2 and 3 drawings, prompting this interchange between the instructor and the interviewees:

> Another difference that I saw when looking at this last set of drawings is when they put children in the classroom they tried to emphasize cultural differences ... Go back and look

Table 3.2 Correlations among student affect codes and pedagogy

	Student neutral	Student positive	Student negative	Traditional instruction	Collaborative instruction
Student neutral	1	−0.042	0.046	0.151	−0.124
Student positive	−0.042	1	−0.099	−0.005	0.229*
Student negative	0.046	−0.099	1	0.331**	−0.222*
Traditional instruction	0.151	−0.005	0.331**	1	−0.257*
Collaborative instruction	−0.124	0.229*	−0.222*	−0.257*	1

N = 112

*Correlation is significant at the 0.05 level (2-tailed)

**Correlation is significant at the 0.01 level (2-tailed)

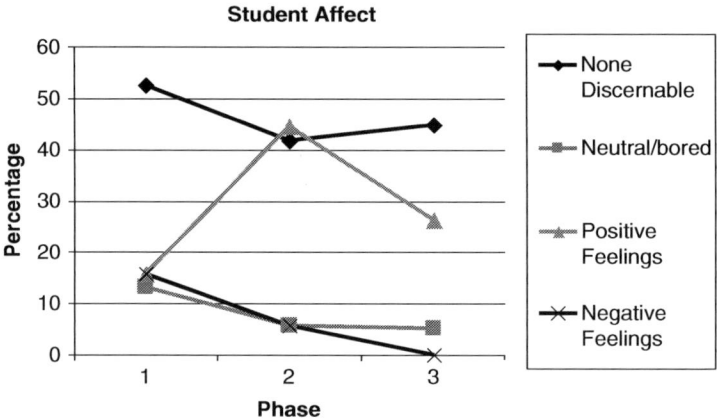

Fig. 3.6 Student affect from phases 1 through 3

at the drawings … You're right! I never even noticed that. In a lot of them, if they had students in the classroom, you can start to see the difference. Another thing they did, if they placed students in the classroom, they showed gender.

Illustrating students with distinctive features humanizes them. It represents a general consensus among the prospective teachers that individual students possess individual learning needs, which suggests an increase in the likelihood that teachers would respond to students as distinct persons and not merely grouping them as interchangeable objects of instruction.

Trends in Mathematical Representations and Content

A finding associated with changes in pedagogy was the decreasing trend in abstract representations along with an associated upward trend in concrete representations (Fig. 3.7). While almost 90% of Phase 1 drawings showed abstract representations,

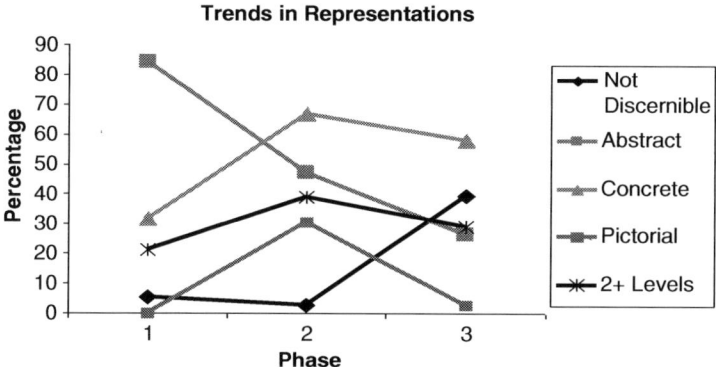

Fig. 3.7 Trends in representations from phases 1 through 3

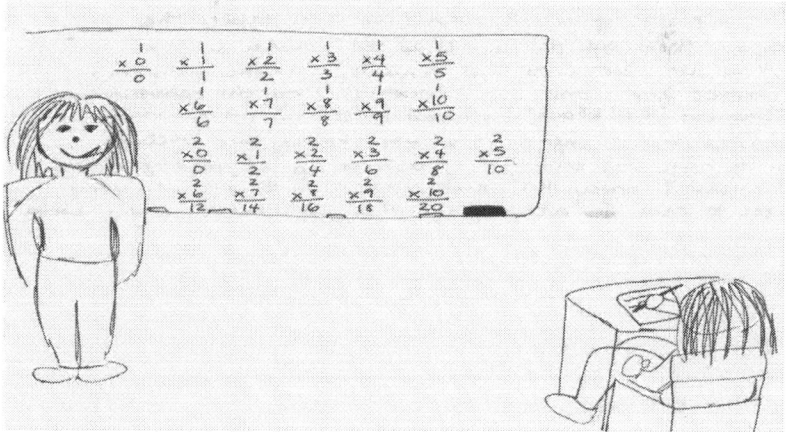

Fig. 3.8 Abstract representations in phase 1 (multiplication exercises)

Drawings in Phases 2 and 3 showed more concrete representations. In addition, drawings depicting more than one representation increased slightly from Phase 1 to Phase 2. The lack of specificity in student affect noted in Phase 3 drawings carried over to a sizeable percentage (40%) depicting no representations. The typical abstract representation depicted in Phase 1 was multiplication problems written on the chalkboard as shown in Fig. 3.8. In contrast, many drawings from Phase 2, such as that in Fig. 3.9, show two or more representations of the same concept: concrete, pictorial, and abstract. In this lesson, students are designing toys with multilink cubes by both building the toys and writing directions for other students to create the same toys.

These drawings also underscore a major difference between the drawings of Phases 1 and 2 – differences in content. As shown in Table 3.3, multiplication was the dominant content portrayed in the drawings from Phase 1, while a greater variety

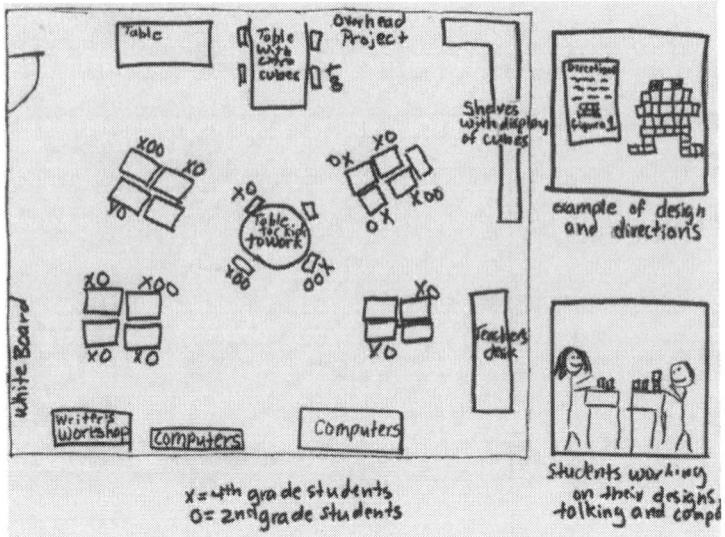

Fig. 3.9 Using manipulatives in phase 2

Table 3.3 Mathematics lesson content noted in phases 1, 2, and 3

Content	Phase 1	Phase 2	Phase 3
None observed	8	0	42
Addition	11	22	8
Classification/sorting	0	25	8
Counting	3	22	3
Decimals	5	0	0
Division	11	11	3
Fractions	5	8	8
Geometry	0	25	3
Graphing	0	8	8
Money	5	6	3
Multiplication	63	17	5
Patterning	0	25	3
Place value/number	5	8	3
Subtraction	8	3	8
Unclear math content	8	8	24
Total	132	189	126

of content was shown in the drawings from Phase 2. Lessons with more than one type of mathematics content depicted per drawing also increased in Phase 2 – as noted in Table 3.3 by the totals for coded content of each Phase. The increases in mathematics content did not carry through to Phase 3 (total for Phase 3 coded content = 61). This fact coincides with the lack of other details in Phase 3 drawings (fewer students and representations presented than in first two Phases, for example).

These aspects of Phase 3 drawings suggest that these prospective teachers' final drawings were of more general mathematics situations – again, underscoring the differences between *Apprenticing* experience, which is a direct process and *Actualizing* experience, which is a more amorphous process. However, Phase 3 *Actualizing* drawings were rich in room details, the implications of which are discussed next.

Changes in Room Structure

Besides changes in pedagogy, the area of most notable change was that of room structure. The prospective teachers emphasized room structure, particularly in the Phase 3 drawings. In these latter drawings, room elements represented particular types of mathematics instruction. For example, rooms with centers allow students to work on different topic areas at the same time; similarly, rooms with tables instead of desks encourage students to work collaboratively.

In an effort to make sense of the many codes related to room structure, composite variables were again created. Room details were categorized as either "traditional" or "collaborative." The Traditional Structure variable tended to coincide with Traditional Instruction. In these rooms, students worked individually while receiving group instruction from teachers. Only the following two codes fit well into this composite: Desks in rows and Minimal furnishings. The variable Collaborative Structure related to Collaborative Instruction – with active and cooperative learning. Here, the following five codes are relevant: Desks in groups, Tables, Centers, Worktables, and Manipulative cart/shelf. Figure 3.10 shows the trends in these variables.

The trends in room furnishings and structures changed dramatically and consistently across the phases. Standardized counts for a Traditional Structure in Phase 1 reached 57, yet for Phase 3 the count was a mere 7. By contrast, standardized counts for Collaborative Structure started at 14 for Phase 1, but reached 42 at Phase 3. This pattern indicates a distinct move away from a classroom structure that supports the teacher transmission model to one in which students work together and work actively. Here, student engagement is valued while passive learning is avoided. A comment from one interviewee illustrates this idea:

> [One] thing that really jumps out at me is that there seems to be more student-to-student interaction than before. Also you don't see the "u" shaped classrooms like before – they are clusters of students sitting together. The other classrooms were clearly set up – there weren't clear indications of students-to-student interactions – [even if students were sitting next to each others] students were still working on worksheets. I think that the physical arrangement of the classroom told a different story this time. Another thing that jumped out was I saw the use of more concrete materials with these.

The increase in the use of technologies and manipulatives was noted when looking at the smaller elements of the drawings. Two more composite variables were then created. Technology is comprised of the following four coded categories: Computer, Calculator, Television, and Cassette Recorder/Player. Manipulatives is a sum of all of

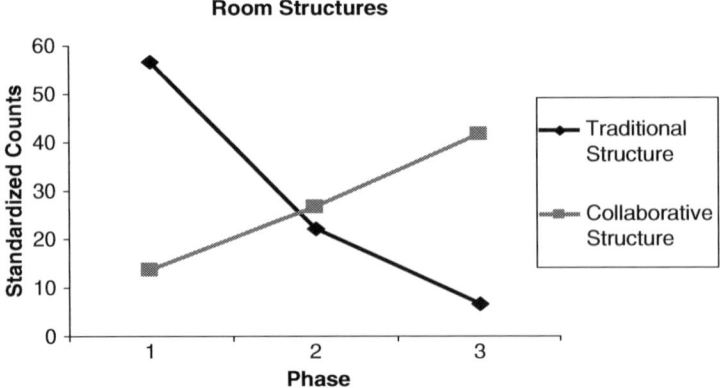

Fig. 3.10 Traditional versus collaborative instruction (room structure)

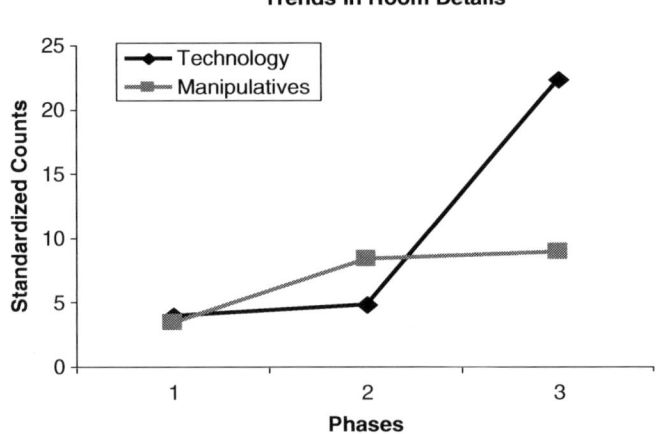

Fig. 3.11 Trends in room detail variables from phases 1 through 3

the coded manipulatives with the exclusion of flashcards and workbooks. A dramatic change was noted in the increase of Technology from Phases 2–3 (from a standardized count of 5–22, respectively). The depiction of computers, in particular, increased markedly from 11% of the drawings in Phase 1, to 17% of the drawings in Phase 2, to 74% of the drawings in Phase 3 (the change between phases 2 and 3 indicating that prospective teachers planned to rely on computers more in their instructional practice than was the case in the practica). As shown in Fig. 3.11, the coding of Manipulatives also increased over the Phases, but this growth wasn't as dramatic – from a standard count of 3 in Phase 1 to a standard count of 9 in Phase 3.

Discussion of Results

Drawing studies allow us to see situations through the perspectives of the participants. This study provided an opportunity to understand teaching practices through the eyes of prospective teachers. The teaching situations indicate that past, present, and future classroom situations are influenced by the sociocultural context in which these perspective teachers experienced mathematics (See Fig. 3.1). The drawings not only capture glimpses of teaching practice from different time periods (where we see the past in Phase 1, the present in Phase 2, and the idealized not-too-distant future in Phase 3), but they also correspond to various levels of power (from the relatively powerless student to the fully vested teacher) as well as varying levels of pedagogical knowledge in the drawing's creator. All of these conditions influence the drawn representations and narratives depicting classroom practice. An additional source of variability was the change in the drawing prompts. All these sources of inconsistency render the trends found herein somewhat precarious, although the systematicity of these trends suggests that some fundamental changes occurred. For example, some perspective teachers indicated that the Phase 1 drawings were heavily influenced by the prompt – depictions of the past. While many prospective teachers created austere images of third grade mathematics learning, students often indicated on the back that they disagreed, in part, with the instruction that was depicted. To explain her illustration (Fig. 3.12), one prospective teacher expressed her reservations this way,

> … I feel that there could have been other ways to teach multiplication to students. Looking back, I realize that just writing on the board didn't help to conceptualize the idea from

Fig. 3.12 Abstract representation in phase 1 (teacher-centered)

addition to multiplication [the concept] that was addressed. The teacher could have used
manipulatives (Cuisenaire rods or physical objects) to represent the same idea. Also it may
help to use a story for some students. I don't believe that writing on the board & [using]
dittos are the best way for students to learn ....

Still, these analyses do point to changes in perception that may be due to
knowledge gained from the mathematics methods course and pre-practicum experiences. Changes that appeared to be tied to classroom work were the increases in
collaborative instruction (particularly indicating scaffolding) and technology in
present and future classroom depictions and the use of manipulatives and concrete
representations in the last set of drawings. The inclusion of mathematics content
other than multiplication (in the second set of drawings) also suggests a greater
awareness of the range of mathematics topics – awareness developed in the course.
Just as Haney et al. (2004) found that student-generated drawings "can provide a
valuable catalyst to document, change, and improve what goes on in schools" (p. 243),
this research found that prospective teacher-generated drawings documented
changes in perception of mathematics teaching and learning over time.

Improvements in mathematics practices documented in the drawings are evident
in today's classrooms – a likely result of education initiatives in the past three
decades. Since the publication of *Everybody Counts, A Report to the Nation on the
Future of Mathematics Education* (1989), American educators have been called to
dramatically alter classroom instruction to raise the standard of achievement for all
students. The National Council of Teachers of Mathematics (1989, 2000) have
advocated for instructional techniques that includes more student engagement and a
more diverse curricula focused on the in-depth learning of content. The Common
Core State Standards (2010) underline practices that effectively integrate subject
matter knowledge with pedagogical content knowledge. International comparative
studies like the Third International Mathematics and Science Survey (TIMSS)
(National Center for Education Statistics, 2002) indicate the pressing need to
improve curricula and instruction in America's mathematics classrooms – particularly
in the middle and high school years. A supplementary classroom study of TIMSS
(Olson, 1995) demonstrates that traditional pedagogies continue to prevail in
American middle school classrooms while more progressive; social-constructivist
pedagogies are used in Japanese and German classrooms. The prospective teachers
in our study, through their studies and pre-practicum experiences, are becoming
aware of these problems and the efforts to correct them. When asked about the
impact of education reform on teachers, one prospective teacher remarked,

... More demands are being placed on the student. Teachers are trying to think of ways to
accomplish more and to allow the children to learn more things more quickly and I think
that they can do that by using the different techniques ... I have noticed in my practicum
that more demands have been placed on the teacher.

These new attitudes contrast sharply with the early mathematics experiences presented in the first phase of drawings, where again, students were largely disengaged
and learning consisted of the passive acquisition of knowledge.

The last set of drawings, with an emphasis on the structure of the classroom,
documents the prospective teachers' strategies for incorporating the new strategies

and content into their future mathematics classrooms. The Phase 3 drawings reflect the bidirectional relationship that exists between the prospective teachers' experiences in their mathematics methods course and their experiences in their practica. Their practica gave them the opportunity to take the pedagogical techniques and strategies they learned in their methods course and experiment on them under the guidance of their cooperating teachers and university supervisor. These two experiences introduced prospective teachers to three levels of scaffolding. First, in the methods course, the prospective teachers' instructor and peers scaffolded their learning and understanding of how to teach the content to elementary school children. Second, in their practica placement, the cooperating teachers scaffolded their understanding of how to apply the pedagogy they learned in class. Finally, the prospective teachers integrated the ideas learned from the first two levels by representing scaffolded learning in their practica classrooms, where they had a direct influence on student learning.

While these drawings are light on specifics, they are heavy on the pedagogy that these perspective teachers have come to value (for example, using manipulatives, grouping, and working with students in the zone of proximal development). In particular, the drawings overwhelmingly indicate a predilection for student collaboration, interactive teaching, and the support of diverse student learning needs. When looking at the Phase 3 drawings, one interviewee observed, "I can see there is more of an attempt to try to meet individual needs, more than one standard way of teaching math, more individualized [instruction]." Therefore, what is evidenced in the three drawings, and in the third set of drawings in particular, is the influence of the socio-cultural contexts on the development of prospective teachers' perceptions of the teaching and learning of mathematics to students. This construct directly relates to the framework presented in section one of this chapter.

The prospective teachers' desire to improve mathematics instruction is perhaps best appreciated in the context of their less-than-ideal early mathematics experiences. By engaging in the activities from their weekly methods course, prospective teachers with poor early mathematics experiences have been given the opportunity to address early feelings of failure and to understand that perhaps it was not they who failed but rather the instruction that failed them. The three Phases of drawings presented prospective teachers with the opportunity to address the misconceptions they once had about mathematics as a direct result of the way in which it was taught. Understanding these common misconceptions gives prospective teachers the power to ensure that their own students do not develop them. The course content, supplemented by the reflective experiences encased in the drawings, gives these prospective teachers an appreciation of their power to improve mathematics experiences of a new generation of pupils – and some of them have the opportunity to revise their own mathematics histories from mathematics phobic to revolutionary. One interviewee comment captured this objective best after reflecting on the Phase One's drawings, "When I look at the first set of drawings, I think, 'What can I do to change these perceptions?'" In the first set of drawings it emphasizes show sterile and stagnant mathematics can be, how traditional it is, how it focuses on skills, and definitely how the drawings are teacher-centered. Furthermore, the examination of several drawings across time allows prospective teachers the opportunity to make sense of teaching and learning through a sociocultural context or discourse that may inform as well as persuade their thinking and understanding of mathematical pedagogy.

Conclusion

These findings indicate that through the mathematics course experiences and this drawing project, prospective teachers have become aware of the problems of the past and are actively seeking solutions to these problems. Moreover, according to these drawings, the future of mathematics instruction is promising, as illustrated in Fig. 3.13. Richards (2006, 1996) explains that, with respect to prospective teachers, "looking at one's practices in an ongoing, careful, and deliberate way is crucial to professional growth" (p. 4). The reflective practice of drawing their own teaching and learning experiences gives prospective teachers a real chance for authentic professional intellectual growth – to reflect on their own mathematics experiences, to critique them, and to propose improvements and alternatives. Mathematics teacher education programs have a responsibility to ensure that their graduates possess the pedagogical skills needed to develop into effective teachers.

This study reveals why self-reflection through drawings is emerging as a new method of scaffolding important pedagogical skills for perspective teachers.

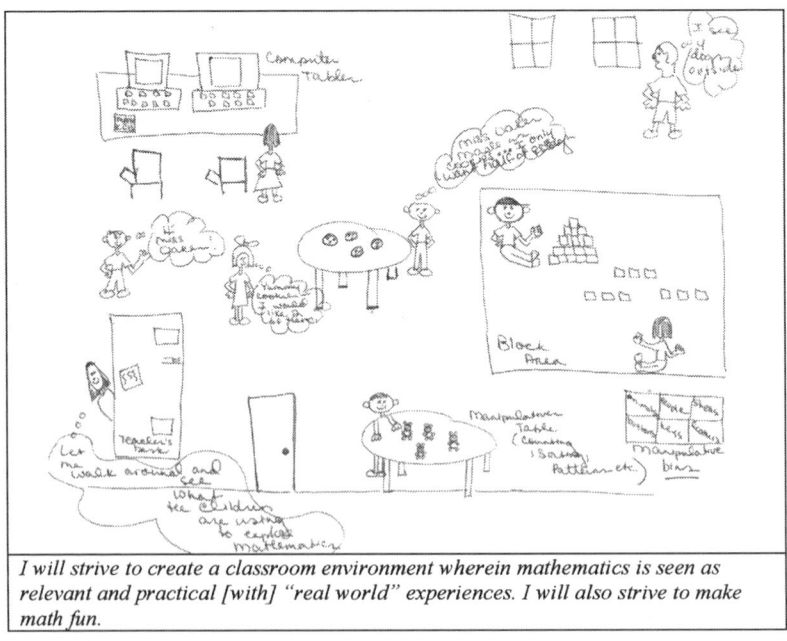

I will strive to create a classroom environment wherein mathematics is seen as relevant and practical [with] "real world" experiences. I will also strive to make math fun.

Fig. 3.13 Drawing from phase 3 (student-centered)

Mathematics teacher educators can use drawings and narratives as a *communicative tool* to assist prospective and practicing teachers in thinking about and questioning the tacit assumptions that underlie their pedagogical practices, rendering those assumptions explicit and open to transformation. This study demonstrates that drawings with accompanying narratives can work to substantially raise the quality of teacher preparation in helping prospective teachers develop positive and actionable conceptions of themselves as mathematics teachers. It is our belief that this work contributes to a contemporary discussion about improving the quality of mathematics teachers and their preparation. This work places teachers' conceptions of themselves and their practices in a broader sociocultural context. As they become aware of their own learning, they are more likely to become aware of the influence of the child's learning context within the mathematics classroom.

Appendix A: Code Definitions for Teacher Drawing Project

Teacher:

Not depicted:	No teacher depicted
1+ depiction of same teacher:	Same teacher shown more than once
1+ teachers:	More than one teacher shown
At blackboard/in front of room:	Teacher is either at the front of the room or at the blackboard
In center of class:	Teacher shown at the center of the class
Instructing class/talking/asking:	Teacher is clearly instructing the whole class, talking to the class, or asking questions of the class
Instructing group/talking/asking:	Teacher is clearly instructing a smaller group of students, can include talking or asking questions
Sitting in chair/instructing:	Teacher is instructing students, while sitting in a chair
Walking toward students/moving:	Teacher is moving toward students or moving around students in class

Teacher affect: Check only if teacher is depicted in drawing or write-up

None discernable:	Cannot make out the teacher's expression
Neutral/bored	
Positive feelings	
Negative feelings:	Sad, scared

Students:

Not depicted:	No students depicted
At blackboard:	Student(s) is/are standing at blackboard
At centers:	Students are shown working at centers
At rug:	Students are working or receiving instruction at a rug
Instructing class/group:	Students are clearly teaching something to the class or a small group
Raising hands:	Students are shown with their hands raised
Sitting in chairs:	Students are shown sitting in chairs without desks/tables

(continued)

Sitting at desks/tables:	Students are sitting at desks or tables
Talking to students:	Students are talking to each other
Talking to teacher/responding:	Students are talking to the teacher or answering a question posed by the teacher
Students moving:	Students are moving about the classroom
Student affect:	**Check only if students are depicted in drawing or write-up**
None discernable:	Cannot make out the students' expressions
Neutral/bored	
Positive feelings	
Negative feelings:	Sad, scared
Desks:	As seen in drawing or indicated in write-up
None drawn	
In groups:	Desks are in clusters – not in rows or lines. Difference between groups of desks and tables is that lines are drawn signifying individual desks
In pairs:	Desks in pairs
In rows:	Desks form a line of sorts – could be a "u" shape – so that students may be seated next to each other but are not facing each other
Tables:	Students are using tables for desks – not to be confused with additional worktables or tables in centers
Furnishings:	**As seen in the drawing or indicated in the write-up**
None depicted	
Minimal:	Usually desks or tables and perhaps a blackboard or wall fixtures
Animals	
Book corner/library:	A place with books for students to go to and read
Centers:	Subject-specific learning areas separate from the desks and tables. If there is a computer center, check this also
Clock/timer:	Some apparatus visible that is keeping track of time
Computer	
Easel	
Manipulative cart/shelves:	Place specifically designated for storage of math manipulatives
Maps	
Meeting rug or area:	Place for students to sit together for instruction or discussion
Musical instrument	
Overhead projector	
Plants	
Record player/CD/tape player, television	
Work table(s)	Tables not designated as centers or desks but as places where small groups can receive instruction
Math content:	Select as many as overtly indicated in drawing or write-up. Do not select content areas related to those obviously indicated

(continued)

(continued)

None observed
Addition, subtraction
Classification/sorting
Counting, decimals, division
Fractions, geometry, graphing
Money, multiplication, patterning
Place value/number sense
Unclear math concept

Tasks/activities:	**Select any that are shown in the drawing or indicated in the write-up**
None observed	
Activity: active	Students engaging in active pursuits beyond writing, listening, reading, answering teacher's questions
Art/music	
Competitive:	Activity in which there are winners and losers – often in the form of a contest or game – may include vying for teacher's attention
Cooperative:	Activity in which two or more people work together to achieve a common goal; students working collaboratively together. Note: a single drawing may be coded as both containing competitive and cooperative task elements as in team competitions
Interdisciplinary:	Any lesson/activity that teaches a mathematical principal through another discipline (e.g., science)
Manipulatives:	Used in the activity for the purpose of teaching math
Paper/pencil:	A paper-and-pencil activity
2+ activities:	More than one activity is going on at once
Individual/seat work:	Individuals working alone on their own work
Group/partner work:	Students are working in groups or in pairs
Whole class activity:	The whole class is working on the same activity/lesson. Teacher is leading the class.

Level of representation:	**Check all indicated in drawing or write-up**
None discernable	
Abstract:	Mathematical concept presented symbolically, orally, can include mental math
Concrete:	Mathematical concept presented with objects
Pictorial:	Mathematical concept presented with pictures, diagrams, and graphs
2+ levels:	More than one level of representation is depicted in the lesson

Manipulatives:	**Check any that are depicted in the drawing or write-up**

None discernable
Base-ten-blocks
Blocks (generic), shape blocks (attribute, tangrams, pattern blocks)
Calculators
Clay

(continued)

Counters/household objects	
Cubes (unifix, multilink, other connecting cubes)	
Flashcards	
Food/candy	
Games/bingo	
Money/coins	
Paper manipulatives:	Looks like manipulatives but made of paper (e.g. pattern blocks made of paper)
Science objects	
Workbooks/worksheets	
Other	

References

Albert, L. R. (2000, April). *A Vygotskian framework for teachers as mathematical problem solvers.* Paper presentation at the American Educational Research Association Annual Meeting, New Orleans, Louisiana.

Albert, L. R., & Rhodes, K. (2005). Prospective teachers' perception of teaching and learning mathematics through images and drawings. In G. M. Lloyd, M. R. Wilson, J. L. M. Wilkins, & S. L. Behm (Eds.), *Proceedings of the 27th annual meeting of the North American Chapter of the International Group for the Psychology of Mathematics Education* [CD-ROM]. Eugene, OR: All Academic.

Black, K. (1991). How students see their writing: A visual representation of literacy. *Journal of Reading, 3*, 206–214.

Bruner, J. (1996). *The culture of education.* Cambridge, MA: Harvard University Press.

Bullough, R. V., Knowles, J. G., & Crow, N. A. (1991). *Emerging as a teacher.* New York: Routledge.

Common Core State Standards Mathematics Initiative. (2010). *Common core state standards: Application of common core state standards for English language learners.* Retrieved June 6, 2010, from http://www.corestandards.org

Dickmeyer, N. (1989). Metaphor, model, and theory in education research. *Teachers College Record, 91*, 151–160.

Elbaz, F. (1991). Research on teacher's knowledge: The evolution of a discourse. *Journal of Curriculum Studies, 23*(1), 1–19.

Goodenough, F. L. (1926). *Measurement of intelligence by drawings.* Chicago: World Book Company.

Gulek, C. (1999, April). *Using multiple means of inquiry to gain insight into classrooms: A multi-method approach.* Paper presented at the meeting of the American Education Research Association, Montreal, Canada.

Haney, W., & Gulek, C. (1996). *Technical manual on student reflection survey.* Unpublished manuscript, Boston College.

Haney, W., Russell, M., & Bebell, D. (2004). Drawing on education: Using drawings to document schooling and support change. *Harvard Education Review, 74*, 241–272.

Haney, W., Russell, M., Cengiz, G., & Fierros, E. (1998). Drawing on education: Using student drawings to promote middle school improvement. *Schools in the Middle: Theory and Practice, 6*(5), 38–43.

Hibbing, A., & Rankin-Erickson, J. (2003). A picture is worth a thousand words: Using visual images to improve comprehension for middle school struggling readers. *The Reading Teacher, 8*, 758–769.

Hill, H. C. (2004). Professional development standards and practices in elementary school mathematics. *The Elementary School Journal, 104*, 345–363.

Kendrick, M., & McKay, R. (2001). Drawings as an alternative way of understanding young children's constructions of literacy. *Journal of Early Childhood Literacy, 4*(1), 109–128.

Krueger, R. A. (1994). *Focus groups: A practical guide for applied research.* Thousand Oaks, CA: SAGE Publications.

Kvalseth, T. O. (1991). A coefficient of agreement for nominal scales: An asymmetric version of Kappa. *Educational and Psychological Measurement, 51*(1), 95–101.

Malchiodi, C. A. (1998). *Understanding children's drawings.* New York: The Guilford Press.

Manning, B. H., & Payne, B. D. (1993). A Vygotskian-based theory of teacher cognition: Toward the acquisition of mental reflection and self-regulation. *Teaching and Teacher Education, 9*, 361–371.

Mason, C. L., Kahle, J. B., & Gardener, A. L. (1991). Draw-a-scientist test: Future implications. *School Science and Mathematics, 91*(5), 193–198.

Miller, S. I., & Fredericks, M. (1988). Uses of metaphor: A qualitative case study. *Qualitative Studies in Education, 1*(3), 263–272.

National Center for Education Statistics. (2002). *Third international math and science survey (TIMSS).* Retrieved May 14, 2002, from the World Wide Web: http://www.http://nces.ed.gov/timss/timss95/index.asp

National Council of Teachers of Mathematics. (1989). *Curriculum and evaluation standards for school mathematics.* Reston, VA: Author.

National Council of Teachers of Mathematics. (2000). *Principals and standards for school mathematics.* Reston, VA: Author.

National Research Council. (1989). *Everybody counts. A report to the nation on the future of mathematics education.* Washington, DC: National Academy Press.

Olson, L. (1995). School portraits: When it comes to sizing up what students think about education, a picture may be worth a thousand words. *Education Week*, 29–30.

Riessman, C. K. (1993). *Narrative analysis.* Newbury Park, CA: SAGE Publications.

Richards, J. C. (1996). Creating self-portraits of teaching practices. *The Reading Professor, 18*, 4–19.

Richards, J. C. (2006). Post modern image-based research: An innovative data collection method for illuminating preservice teachers developing perceptions in field-base courses. *The Qualitative Report, 11*, 37–54.

Scott, L. M. (1994). Images in advertising: The need for a theory of visual rhetoric. *Journal of Consumer Research, 21*, 251–273.

Seidman, I. E. (1991). *Interviewing as qualitative research: A guide for researchers in education and the social sciences.* New York: Teachers College Press.

Tovey, R. (1996, November/December). Getting kids into the picture: Student drawings help teachers see themselves more clearly. *Harvard Education Letter, 6*, 5–6.

Uebersax, J. (2002, July 20). *Kappa coefficients.* Retrieved May 10, 2006, from http://ourworld.compuserve.com/homepages/jsuebersax/kappa.html

Vygotsky, L. S. (1978). *Mind in society: The development of higher psychological processes.* Cambridge, MA: Harvard University.

Weber, S., & Mitchell, C. (1995). *That's funny, you don't look like a teacher: Interrogating images and identity in popular culture.* Washington, DC: Falmer Press.

Weber, S., & Mitchell, C. (1996). Drawing ourselves into teaching: Studying the images that shape and distort teacher education. *Teaching and Teacher Education, 12*(3), 303–313.

Wheelock, A., Bebell, D. J., & Haney, W. (2000, November 2). What can student drawings tell us about themselves as test-takers in Massachusetts? *Teacher College Record.* Retrieved March 20, 2006, from http://www.tcrecord.org

Chapter 4
Improving Teachers' Mathematical Content Knowledge Through Scaffolded Instruction

Introduction

Scaffolds are the supports provided by knowledgeable others to help a learner move from a current level of performance to a more advanced level. Essential to scaffolding within instruction is the use of language for mediation (Albert, 2000; Wertsch, 1980). Language provides the medium for interchange between the scaffolder and the learner, allowing for individual construction and co-construction of knowledge. This study explores the language of scaffolding among practicing middle school mathematics teachers during a professional development program designed to increase mathematical content knowledge. We sought to investigate how the dialogue and interactions among the teachers improved their content knowledge.

This chapter begins by sketching the relevant conceptual aspects of scaffolding as they apply to group interaction. The focus here is on how language and its role in scaffolded learning serve as conduits for investigating learners' thought processes through collaborative problem solving. The next section describes the methods employed, as well as the context for this inquiry. It consists of a description of the salient structure of the professional development seminars and the complex aspects of the group-members' relationship to each other and the mathematical content of the seminars. We then present an analysis of findings that includes the participants' dialogue while learning in collaborative groups in which we discuss the challenges faced by the participants in the seminars as well as the growth that occurred as they engaged in collaborative problem solving activities. We conclude this chapter with implications that scaffolded activities can bring middle school teachers to deeper levels of mathematical understanding.

L.R. Albert, *Rhetorical Ways of Thinking: Vygotskian Theory and Mathematical Learning*, DOI 10.1007/978-94-007-4065-5_4,
© Springer Science+Business Media Dordrecht 2012

Theoretical Framework and Relevant Research

A sociocultural historic model of learning (Cobb, 1994; Goos, 1999, 2004, 2005; Vygotsky, 1978, 1994) provides a theoretical framework for the idea that all higher order processes are first social processes, in which social interactions are key to the learner's development. Vygotsky (1978) writes, "learning awakens a variety of developmental processes that are able to operate only when [learners] are interacting with people in [their] environment and in cooperation with [their] peers" (p. 90). Through interaction, learning takes place within a learner's zone of proximal development (ZPD) (Vygotsky, 1978); "[l]earning within the ZPD occurs when students are involved with tasks or problems that go beyond their immediate individual capabilities in which teachers assist their performance or in collaboration with more knowledgeable peers" (Albert, 2000, p. 109). Within a learner's ZPD, scaffolds function as tools that provide support and allow for the completion of tasks not otherwise possible by the less knowledgeable peer.

The process of scaffolding enables learners "to solve a problem, carry out a task, or achieve a goal which would be beyond [their] unassisted efforts" (Wood, Bruner, & Ross, 1976, p. 90). Effective scaffolding awakens a "variety of internal developmental processes that would be impossible apart from learning," thereby enhancing independent thinking and problem solving (Vygotsky, 1978, p. 90). As learners' independent thinking and problem solving processes are internalized, the supports that scaffold learning fade out and are replaced with more sophisticated structures (e.g., through the learning of algebraic concepts, students move towards understanding mathematics operations as specific examples that support algebraic operations). The underlying premise is the notion that scaffolding is not a unilateral process, but rather a co-constructed one (Rojas-Drummond, 2000).

This scaffolding process implies a social system in which learners' dialogue and interactions with knowledgeable peers actively construct and support knowledge (Albert, Mayotte, & Phelan, 2004). Within a group situation, the same individual does not always assume the role of the more knowledgeable other. Depending upon the demands of the task and the level of interactions among the participants, learners have the "opportunity to shift in and out as the more knowledgeable other when it is appropriate according to individual understandings of the task at hand" (Albert & McKee, 2001, p. 16). Explicit to this process is the Vygotskian view that learning and understanding occur on the following two distinct levels: interpsychological and intrapsychological – between people and then inside the learner, respectively. Furthermore, research confirms that language mediation in social contexts assists learners in their development of higher cognitive processes (Albert, 2000; Doolittle, 1997; Palinscar, 1986; Palinscar & Brown, 1988; Rosenshine & Meister, 1992; Vygotsky, 1978; Wegerif & Mercer, 2000; Wertsch, 1979).

A number of researchers have explored the role of language use in scaffolded instruction (Hogan & Pressley, 1997; Larkin, 2001; Murray & McPherson, 2006; Osana & Folger, 2000; Palinscar, 1986; Palinscar & Brown, 1988; Roehler & Cantlon, 1997). One of the earlier studies was conducted in a first grade classroom by Palinscar; the research examined reciprocal teaching, a dialogue strategy that aids reading instruction in which teachers and students jointly participate in a process of questioning, summarizing, clarifying, and predicting. In a study conducted a few years later, Palinscar and Brown explored reciprocal teaching through collaborative problem solving. These collaborative activities engaged remedial and special education students in critical thinking during reading and listening comprehension activities. Their research showed that teachers assume a more directive scaffolding approach in the early phases of reciprocal teaching; however, directive scaffolding diminished as learners became more independent in applying strategies that assisted their understanding of the task at hand. In a similar study, Hogan and Pressley investigated the types of statements that teachers used to help facilitate student thinking during whole-class discussions. They concluded that when teachers used specific language to frame problems, to refocus discussions, and to summarize statements, students were moved to a deeper understanding of the content.

Language is a principal communicative tool for effective group interactions and discourse through which learners' express their understanding of concepts and ideas (Kumpulainen & Mutanen, 2000; Osana & Folger, 2000). Researchers have come to this conclusion after completing a comprehensive discourse analysis of the form, content, and context associated with the dialogue of middle school students working in their social studies and mathematics classes. For the construction of common knowledge, Osana and Folger found that students' talk coalesced into three significant categories: questioning, responding, and debating. For example, a group member would use questions to gain information from another group member. Students' responses or reactions to questions posed seemed to provide explanations, add information, or offer original suggestions in which the dynamic nature of the discourse encouraged group members to debate ideas and beliefs as they worked toward a shared understanding of the concepts under study. Kumpulainen and Mutanen performed a functional analysis of verbal interactions, examined cognitive processing regarding strategies and procedures, and explored social processing regarding types and forms of participation. These researchers categorized functions of language in peer engagements to include the following: way of thinking, evaluative, interrogative, responsive, organizational, argumental, and experiential.

The process of learning from interacting with others is one of the essential aspects of collaborative learning in social contexts. The nature of learning is such that people do not learn in a vacuum; socialization or interpersonal interactions must precede the intrapersonal aspects of development. The contributions of the research presented emphasize the role of language in student-to-student and teacher-to-student interactions in social contexts. What distinguishes the present study is the examination of teacher-to-teacher scaffolding during mathematics professional development activities in which the primary focus is on the role of language situated in collaborative tasks. We investigate how the dialogue and interactions among the teachers improved their content knowledge. Furthermore, the work presented here put forward an example of a practical application of Vygotsky's sociocultural historic theory. The framework presented in Fig. 4.1 provides a conceptualization for the research study presented and the analysis that follows.

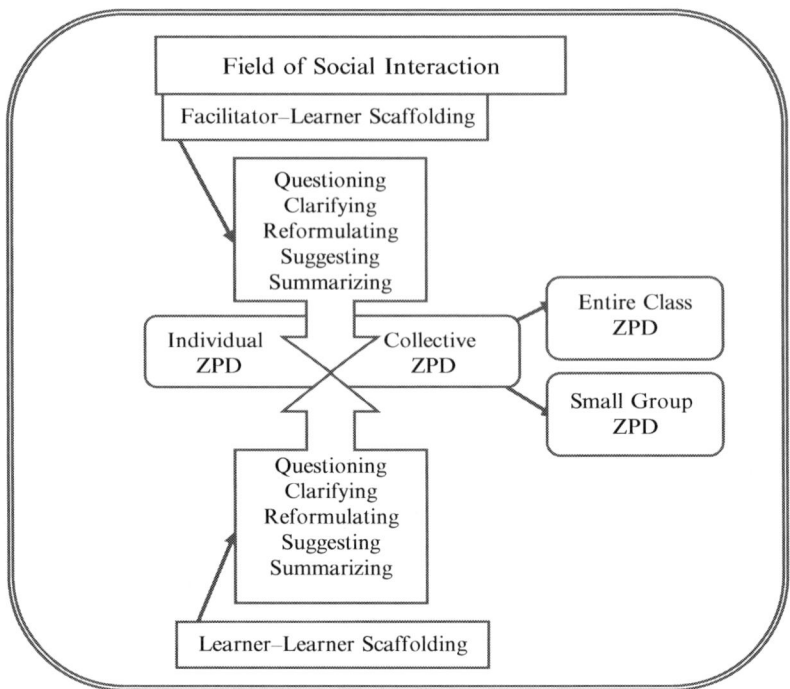

Fig. 4.1 Theoretical and analytical framework

Explanation of the Framework

The *Field of Social Interaction* encompasses facilitator-learner scaffolding as well as learner-to-learner scaffolding within entire class and small group contexts. The entire class dynamic considers the engagement between the facilitator and the learner as new material is presented and initial practice takes place, whereas the small group takes into account the engagement among learners as they grapple with new problems that emerge. The facilitator contributes to the small group context by presenting a probing question or clarifying statement, just as learners contribute to each other through their questions and comments in the large class context. The *Field of Social Interaction* situates individual learners and groups of learners in collaborative processes that scaffold the social and cognitive development of the learners.

In this framework, categories of engagement are operationalized according to the purpose served in the context of entire class and small group work. The five categories identified – questioning, clarifying, reformulating, suggesting, and summarizing – are scaffolds that promote thinking, reflecting, and learning by those engaged in the group interactions. *Questioning* may serve as a clarifying or probing function. Through *clarifying* questions, learners strive to dispel their confusion about previously discussed material. In a *clarifying* response, the scaffolder provides explanation so that the material is easier to understand. *Probing* questions attempt to scaffold the learner to think about the material and/or process in ways not considered initially. *Reformulating* (i.e., rephrasing), another engagement category, allows the learner or the facilitator to verbalize a previously presented thought. *Suggesting*, as an engagement category, offers an opportunity for learners to present ideas for consideration that potentially contribute to others' understanding of the content and may improve group dynamics. *Summarizing* brings together, in a cohesive and constructive way, the important ideas shared within the group. For the present study, we employed these five engagement categories to scaffold the dialogue between and interactions among the teachers as they worked collaboratively to improve their content and pedagogical knowledge.

Methodology

This study used mixed-methods, which merged the strengths of quantitative and qualitative paradigms. The National Research Council (2002) suggests that investigations may be bolstered considerably by using several approaches that incorporate "quantitative estimates of population characteristics and qualitative studies of localized context" (p. 108). According to Johnson and Onwuegbuzie (2004), "[m]ixed methods research offers great promise for practicing researchers who would like to see methodologies described and develop techniques that are closer to what researchers actually use in practice" (p. 15). The first approach was used sequentially to inform the second approach, from which contradictions and new perspectives emerged (Creswell, 2003; Creswell & Plano Clark, 2006; Creswell, Plano Clark, Guttmann, & Hanson, 2003; Greene, Caracelli, & Graham, 1989). The first approach applied a pre-posttest design that included the following: the use of pretests and posttests to assess participants' increased content knowledge, and the use of presurveys and postsurveys with open-ended reflective questions to assess participants' experiences and beliefs about collaborative inquiry.

The second method was an interpretive approach, which simultaneously interwove data collection, data analysis, and data interpretation (Andrade, 2009; Janesick, 1994; Thorne, Kirkham, & O'Flynn-Magee, 2004; Wasser & Bresler, 1996). In particular, this approach was uniquely suited to minimize the impact of the research on participants while maximizing the dialogue of scaffolding and interactions during the mathematics professional development seminars for subsequent analysis and interpretation. An interpretative approach presumes that "reality is socially constructed and the researcher becomes the vehicle by which this reality is revealed ... [and it] is consistent with the construction of the social world characterized by interaction between the researcher and the participants" (Wasser & Bresler, p. 42) Vygotsky (1978) argued that research should result in a dynamic analysis in which "the complex reaction must be studied as a living process, not as an object" (p. 69). Thus, it was important to study processes leading to outcomes. In this case, the focus was on the use of language by the participants to provide information about how a social practice, a practice different from the classroom social context, informed mathematics performance and understanding.

Context and Participants

The professional development context for this research utilized sociocultural practices as originally advocated by Vygotsky (1978, 1994) and later by the work of Davydov (1990, 1995, 1998), Goos (1999, 2004, 2005), Kozulin (1998), and Wells (1999, 2000). Their research suggests that social practices need to be developed not only to engage learners in activities in which they acquire knowledge, but also to engage them in activities that further their intellectual development. Therefore, during the professional development seminars, opportunities were provided for social interaction to benefit the participants' cognitive and social development (Cohen, 1994; Jennings & Di, 1996; Rosenshine & Meister, 1992).

Research further illustrates that teachers' knowledge and classroom practices are readily influenced by professional development that focuses on content knowledge and active learning (Garet, Porter, Desimore, Birman, & Yoon, 2001; Hill & Ball, 2004). "Teaching mathematics requires an appreciations of mathematical reasoning, understanding the meaning of mathematical ideas and procedures, and knowing how ideas and procedures connect" (Hill & Ball, p. 331). Then, an essential way to influence the teaching of mathematics in classrooms is through quality professional development activities, which focus on mathematical knowledge for teaching (Cohen & Hill, 2001; Hill, 2004). As a consequence, the professional development needs of the teachers, the cognitive and social gains as a result of collaborative group work, and the use of quality open-ended tasks contributed to the mathematical learning process of the participants (Cohen, 1994; Osana & Folger, 2000).

The setting for this study was a learning community of middle school mathematics teachers (grades 5–8) serving urban, ethnically diverse, and low-income student populations within Catholic and private schools located in the Northeast. Invitational letters seeking middle grade mathematics teachers interested in professional development and collaborative inquiry related to mathematical problem solving were sent to Catholic and private schools as described above. Teachers interested in participating responded to the researchers and the mathematics learning community was established. Twenty-two teachers volunteered to participate in this study. Of the teachers who participated, the majority were white and female with both general and mathematics teaching experience ranging from a few years (<4 years) to many years (>10 years); the greatest percentage of participants had fewer than 4 years of experience, and those with more than 10 years of experience comprised a much smaller percentage of the total. Half of the participants held graduate degrees either in mathematics education or middle school teaching (50%). Table 4.1 presents a summary of the descriptive characteristics about the teachers that participated in this study.

Data Sources and Procedures

The results that are presented in this chapter come from the following data sources: a pretest and posttest for content knowledge, a presurvey and postsurvey comprised of open-ended reflective questions, transcripts of audio recordings of seminar dialogue

Table 4.1 Participant characteristics

Gender		Race		Degree		Teaching experience		
							General	Math
Male	27%	Asian	5%	Undergraduate	50%	<4 years	45%	55%
		Black	9%			4–10 years	22%	32%
Female	73%	Latina	9%	Graduate	50%	>10 years	32%	14%
		White	77%					

Note: $N = 22$ (participants)

and activities, and observation field notes. Parallel forms of the pretest and posttest focused on five content standards: number and operations; algebra and functions; data analysis; measurement; and probability and statistics. The pretest and posttest were based on released items selected from the Eighth and Tenth Grades 2002 and 2003 Comprehensive Assessment System (CAS) Examinations.[1] The CAS Examinations were selected because the participants were not familiar with these test items given that they all taught in Catholic or private schools that were not required to administer the state sponsored examination. The pretest and posttest consisted of six multiple-choice items, two short answer items, and two open-response items. The short-answer items were problems that relied on basic algorithmic solutions such as "Let x be a positive even number that is less than 10. Write one ordered paired (x, y) that would make the equation $y = x + 3$ true." Open response items were multi-steps problems that went beyond applying a basic algorithm, which required written explanations to illustrate thinking and understanding. For the purposes of this study, one open-response item was constructed as a group task to be completed with participants working in pairs or small groups of three or more members. A graduate assistant with a masters degree in mathematics scored the pretest and posttest that included using a rubric to score the open-response items.

The purpose of the presurvey and postsurvey was to assess the participants' beliefs about and experiences with collaborative inquiry. For the survey, collaborative inquiry was defined as a process of interaction with other teachers during which participants analyzed questions and problems and worked together to formulate appropriate answers and solutions for the purpose of instructional improvement. For example, participants were asked to describe an experience of collaborative inquiry they had been involved with prior to their participation in the study. Another question asked, "Do you prefer to work individually or collaboratively as you engage in mathematical problem solving? Provide an explanation for your preference." The professional development seminars emphasized content and instructional strategies in a collaborative environment in which each seminar lasted for approximately 2 h. Observation techniques and audiotaping of group engagement were used to document this process. Mathematical topics covered in the seminars included

---

[1] The actual name of the state comprehensive assessment test is not given because to name the test would provide some identification of the Catholic archdiocese district. This might lead to identification of schools and teachers and would violate the human subject agreement with participants that all identifying information would be excluded in dissemination of results.

the following: basic concepts of number and number operations, measurement, data analysis, and patterns, functions and algebra. Problem solving and mathematical reasoning skills were emphasized within each topic development.

The facilitators established the tone for a relaxed seminar by welcoming participants and providing refreshments and convivial surroundings. The pretest and posttest measures were completed in the first and last seminars, including the completion of demographic information. Each seminar meeting was tape-recorded and later transcribed by a research assistant. In addition, two facilitators were always present at the seminars, with one of the facilitators serving as an ethnographer, taking detailed observation notes to document the leading facilitator's interactions with participants. The notes were incorporated into the transcripts for analysis. Participants signed informed consent forms, which included notification that responses would be anonymous and the tapes would be destroyed after completion of the study.

At the seminars' onset, participants experienced the learning and understanding of mathematics through the collaborative group approach known as "complex instruction." Complex instruction (CI) is a collaborative group strategy in which individuals work together in groups so that everyone can participate and work on a collective task that has been clearly designed (Cohen, 1994). The assumption is that *learning and knowledge construction* develops with assistance from a more knowledgeable other, focusing on learning as an unfolding process, measured moment-by-moment. The remaining seminars focused on content and pedagogical knowledge. During this process, the participants worked unassisted as they engaged in whole group activities led by a facilitator. Next, the participants worked in a collaborative context on a similar but more complex problem solving activity. Interpersonal interaction provided a context in which the tool of language (talk) functioned as a bridge between the activity and the participant's prior knowledge and thereby assisted the learning process. Table 4.2 presents a summary of content and activities for the professional development seminars.

Data Analysis

This study utilized a mixed-method technique, applying both quantitative and qualitative analyses. Quantitative analysis involved descriptive statistics to indicate the mean score and the variability of scores for the sample of participants for the content knowledge pretest and posttest measures (using SPSS 11). Next, applying inferential statistics, data generated from the content pretest and posttest were analyzed by comparing whether the observed measure of pre-post gains differed significantly, while simultaneously applying a t-test for correlated means. As the same participants took both the pretest and posttest, it seemed appropriate to apply the t-test for correlated means rather then the t-test for independent means (Gall, Gall, & Borg, 2010; Gay & Airasian, 1996).

Explanation and interpretation of qualitative data generated from the seminar activities clustered around the two following broad categories: understanding participants' learning of mathematical content and understanding the role of language

Table 4.2 Professional development seminars

Seminar topic	Length (hours)	Sample activity
Complex instruction: a collaborative group approach	3	Content pretest
		Presurvey
		Introduction to complex instruction (CI)
Number and operations: whole numbers	2.5	Brief review of CI
		Explorations of number operations with base-ten-blocks
		Small group and paired problem solving using CI strategies
Number and operations: fractions and decimal algorithms	2.5	Exploration of fraction algorithms with pattern blocks
		Small group and paired problem solving
Algebra, patterns, and function: part I	2.5	Activities with balance scales, Curisenaire rods, and graphic representations
		Small group and paired problem solving
Algebra, patterns, and function: part II	2	Activities with pattern blocks, cubes, and algebra tiles
		Small group and paired problem solving
Probability	2	Simulation activities with concrete objects: theoretical and experimental probability
Complex instruction: follow-up	2	Review and introduce new aspects of CI
		Small group activity
Using technology	2	Technology as an instructional resource
Wrap-up	1	Content posttest
		Postsurvey

for teaching that content. These categories emerged from the predefined framework (i.e., *Field of Social Interaction*) discussed in an earlier section of this chapter. Using the predefined framework, we knew that codes would transform and develop as the seminar experience continued for the participants (Miles & Huberman, 1994; Thorne, Kirkham, & O'Flynn-Magee, 2004). Thus, the unit of analysis for the seminar activities focused on the *talk* that took place as the participants actively engaged in *questioning, clarifying, reformulating, suggesting,* and *summarizing* mathematics content. Talk as the unit of analysis provided a level of focus for a variety of components inclusive in the seminars. For example, how participants responded to the content was considered with regard to participants' interactions with their peers and the facilitator. Additionally, consideration was given to the participants' assessment of involvement in the seminars' tasks. To be consistent with the methodological approach of mixed-methods, data source triangulation was achieved through the use of observations, interview transcripts, and open-ended surveys. As an illustration, employing these multiple data sources helped develop an interpretative stance from which to appreciate the teachers' dialogue about how to scaffold their students' understanding of the relationship between fractions and decimal numbers. This information presented valuable insights into the teachers' thinking and also provided an analysis of conflicts that may arise, highlighting potential areas of conceptual misunderstanding.

Findings

In the following sections, we present the findings that emerged concerning the type of talk that best scaffolds adult learners and moves them forward in their strengthening of mathematical content knowledge. We begin with the results from an analysis of the content knowledge pretest and posttest and the presurvey and postsurvey. Next, we present the results that describe the dialogue of the participants. The findings in this section exemplify how the language of scaffolding among the practicing middle school mathematics teachers during professional development seminars affected their understanding of the content. We use the teachers' exact words to give readers a sense of their voices and to authenticate what scaffolding looks like in a professional development seminar with regards to the participants' thinking, processing, and mathematical reasoning.

Content Knowledge: Pretest and Posttest Results

We ran tests for the four following categories: the individual open-response item, the group open-response item, the total test score, and the multiple-choice section and short-answer items. For the purpose of analysis, the multiple-choice and short-answer items were merged for a single score because each of these items received a raw score of one point. There was improvement from the pretest to posttest scores of all teacher-learners in all four of the categories. However, statistically significant results at the .05 level surfaced only in the group response item, the total test score, and in the multiple-choice items. These results are highlighted in Table 4.3, which includes the minimum and maximum score for each category with mean (M) and standard deviation (SD).

Presurvey and Postsurvey Results

A collection of the findings was formed around survey results involving collaborative inquiry. For example, participants were asked if past professional development experiences could be described as collaborative inquiry: 64% said yes, and 36% said no in presurveys; yet by the postsurvey, 100% described their professional development experiences as incorporating collaborative inquiry. This was not a surprising result because the participants were most likely incorporating the seminars, which were collaboratively structured, in their postsurvey responses.

Based on the results of presurveys and postsurveys, it can be stated that before the professional development seminars, the majority of teachers (64%) preferred to work individually when learning new material and that there was very little change after the completion of the seminars, with 59% still listing their preference

Table 4.3 Pretest and posttest means and standard deviations

Item (range of possible points)	Pretest		Posttest	
	M	SD	M	SD
Multiple choice and short answer (0–8)	5.09	1.15	5.68*	0.56
Individual open- response (0–4)	3.23	1.32	3.77	0.53
Group open response (0–4)	2.91	1.57	3.73*	0.46
Total (0–16)	12.86	3.04	14.50*	1.54

Note: N = 22 (participants)
*Statistically significant score increase

for learning new material at an individual level. However, postsurvey results show that there was an increase in the percentage of participants stating that they preferred learning new material collaboratively; such responses increased from 22% in presurvey responses to 41% in postsurvey responses. Analysis of data for the question that asked participants how they preferred to work when engaging in mathematical problem solving activities illustrates that the preference for working individually shifted from 55% to 36% and the preference for working collaboratively shifted from 41% to 64% from presurveys to postsurveys. Another noteworthy finding is that there was very little change in the amount of hours participants spent working with colleagues to discuss instructional issues, classroom activities, or curriculum. Overall, the amount of time did not change from before the seminars started to after the completions of the seminars, with a slight increase in the amount of time working with colleagues for 5 h or more, from 9% to 14% respectively. Table 4.4 presents a summary of these results.

The presurvey and postsurvey asked the participants to list what they believed to be the three most significant benefits of collaborative inquiry. The top five most frequently stated benefits are listed in Table 4.5, with collegiality and classroom management being listed more frequently than any other benefit on the presurvey. The most frequently listed benefit on the postsurveys focused on the process of interaction and included understanding the importance of multiple perspectives and developing knowledge and understanding of student learning. It is important to note that student learning, affirmation, and improving pedagogical practices were three of the five most commonly listed benefits for both the presurveys and postsurveys. Qualitatively, there were changes in the nature of the responses to this question; responses in the presurveys were typically short phrases, whereas responses in the postsurveys were more likely to provide an explanation, which showed greater depth and reasoning about the role of collaboration in teaching and learning. These explanations communicated sentiments about student learning similar to the response of one participant who said, "I always thought some students needed to think alone. But now I realize that they could learn from each other." Another participant wrote that the seminars helped her learn techniques and skills that "our students need when working with others. Yet, I believe that it is essential to have a sense of balance between collaborative and individual work so that students may learn from each other but are still able to show their individual understanding."

Table 4.4 Participant responses to presurvey and postsurvey

Item	Presurvey	Postsurvey
Past collaborative experiences		
Yes	0.64	1.00
No	0.36	0.00
Learning new material preference		
Individually	0.64	0.59
Collaboratively	0.22	0.41
Both	0.14	0.00
Problem solving preference		
Individually	0.55	0.36
Collaboratively	0.41	0.64
Both	0.09	0.00
Hours spent working with colleagues		
0–1 h	0.60	0.58
2–4 h	0.31	0.28
5 or more hours	0.09	0.14

Note: Results expressed as percent responding

Table 4.5 Top five benefits of collaboration

Presurvey	Postsurvey
1. Collegiality	1. Understanding the importance of multiple perspectives for all group members
2. Classroom management	2. Developing knowledge and understanding of student learning to improve instruction
3. Student learning	3. Affirmation
4. Affirmation	4. Improve pedagogical practices to improve student learning
5. Improve pedagogical practices	5. Knowledge construction of content to improve teaching and learning

Talk and Scaffolded-Interactions

As stated earlier in this chapter, several researchers have made careful investigations of the role of language use in scaffolded instruction, concluding that scaffolding by the teacher in many cases determine the level at which students acquire an understanding of the content (Roehler & Cantlon, 1997; Hogan & Pressley, 1997; Osana & Folger, 2000; Palinscar, 1986; Palinscar & Brown, 1988). Our objective was to investigate how the dialogue among the participants – who at this point will also be referred to as *teacher-learners* – scaffolded their understanding of content knowledge. We were interested in capturing, as well as qualitatively describing, teacher thinking about mathematics concepts that formed the basis of how they teach that content. In order to achieve this aspect of the investigation, we had to involve the teacher-learners in professional development activities that were beyond prior learning experiences. As a starting point, one of the seminar activities

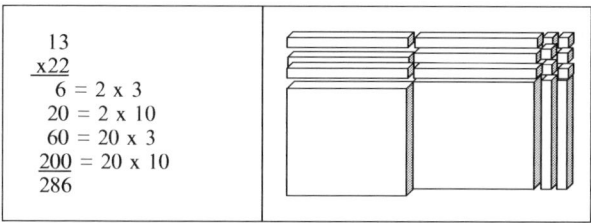

$$
\begin{array}{r}
13 \\
\underline{\times 22} \\
6 = 2 \times 3 \\
20 = 2 \times 10 \\
60 = 20 \times 3 \\
\underline{200} = 20 \times 10 \\
286
\end{array}
$$

Fig. 4.2 Illustration of 13×22 with partial products and with base-ten blocks

asked the teacher-learners to use base-ten-blocks to construct a rectangular array of the problem 13×22, to identify the partial products, and to discuss what the model suggests about mathematical teaching and learning of algorithms (Fig. 4.2).

Analysis of data indicates that not all of the teacher-learners were able to represent accurately the multiplication algorithm of 13×22 as an array interpretation using base-ten-blocks. Furthermore, several of the teacher-learners were not successful in describing the physical attributes of the array model that connected the symbolic representations, whether (a) numerical, as in the length of a side (13), or (b) as they relate to the process of multiplication, which created a level of individual tension and struggle. While their written work provided a tangible product that could be evaluated, verbal interaction among teacher-learners communicated a greater sense of the process and the reasoning that led to what appeared on paper. To illustrate these circumstances, consider the following sequence that occurred as the facilitator guided Andrew and Ellen as they constructed a model of 13×22:

Facilitator: Andrew, I see that you framed 22×13 and solved the paper-and-pencil algorithm. Have you ever used base-ten-blocks before?

Andrew: No, but I like them.

Facilitator: Did you use them when your solved the paper-and-pencil algorithm?

Andrew: Yes and no. You see, I just did what you did, without using the blocks.

Facilitator: I used the blocks when I modeled the problem. Okay, what is here, can you identify the partial products (*referring to his paper-and-pencil algorithm*)?

Andrew: I'm not sure I understand. I see them here on my paper, but I'm not sure where they are located here (*pointing to the concrete model of the solution to the problem*).

Initially, Ellen sat quietly listening and observing the interaction between the facilitator and Andrew. She made note of Andrew's confusion around the partial products and readily volunteered to assist Andrew.

Ellen: I think I can help him. (*Speaking to Andrew*). Okay, we will start from the beginning. Do you agree that this is 13? (*Pointing to the part of the model that represents 13 as well as drawing it out on paper*)

Andrew: Yes.

Ellen: Then, where are 22? Show me (*Andrew points to the blocks that repre-sent 22 and Ellen draws a sketch to represent 22 blocks*).

Ellen: Are you following me? The first partial product is 6. Show me 6. How did we get 6?

Andrew: Yes. The six is here. It's the small cubes. Oh, we got this by multiplying 2×3.

Ellen picked up the six small cubes, showing them to Andrew before replacing them. Again, drawing a representation of the six small cubes. Then she asked Andrew:

Ellen: If we get the next partial product by multiplying 2×10, then where is it shown with the blocks? Are you following me? Show me that you understand.

Andrew: Wow! That would be 20 and the product is the two longs. Now I get it. I can see the other two products. This is different from the way I learned it.

When learners encounter a mathematical problem that is beyond what they are apt to understand on their own, the facilitator or more knowledgeable colleague assists the learners, providing support directly or indirectly through hints, sugges-tions, models, questions, or a combination of these scaffolding techniques. Andrew could frame 22×13 with base-ten-blocks as well as solve the paper-and-pencil algorithm, but when asked by the facilitator to identify the partial products and to relate the concrete model to the paper-and-pencil algorithm, he was not able to make the connection. In this instance, the concrete model posed greater difficulty for him than the abstract algorithm. Ellen scaffolded Andrew's understanding by carefully explaining and modeling how she solved the problem with the base-ten-blocks. She provided a systematic explanation in which she would draw an illustration of the model, point to the concrete representation of the partial products, and then perform the algorithm to show the partial products (See Fig. 4.2). At each step of the problem, she would ask Andrew "Are you following me? If so, show me" to which he would reply, "Yes!" as well as demonstrate the procedure, following Ellen's process to show his understanding.

The scaffolding that occurred within learning conversations between the teacher-learners facilitated their thinking through framing, encouraging, refocusing, and prompting. The dialogue strategies used by Ellen provided a means of ongoing assessment of Andrew's understanding of the content. For Andrew, his prior experi-ences in learning and teaching mathematics may have been based on a symbolic rather than concrete representation of numbers founded only on the rote memorization of rules or formulas. Rote learning may have assisted teacher-learners of mathematics in performing the operations correctly, but in learning to answer the problem correctly, they may have never understood the problem-solving processes of math-ematics. In this example, interaction and language between Ellen and Andrew served as mediators to alter or generate new knowledge, creating the zone of proximal development. The zone of proximal development brings the scaffolder (Ellen) and the learner (Andrew) together with the mathematics content, where Andrew is assisted in acquiring the necessary tools for understanding abstract concepts using concrete models (Kozulin, 1998). Throughout this time, the facilitator monitored the interaction, making note of how Ellen assisted Andrew, who was working to understand how to solve mathematical problems with models outside of his practical experiences. It is important to note that while assisting Andrew, the focus was not only on the dialogue that Ellen used to scaffold his understanding, but also on the concrete actions she performed on the manipulatives. The concrete actions were essential because they grounded the mathematical algorithm that he used initially to solve the problem.

The examination of how the teacher-learners organized their talk and actions and how they scaffolded their learning provided some sense of their pattern of thought, its composition and structure, and the role of the facilitator. When we gave these aspects careful consideration, it became clear that the *Field of Social Interaction* came into play in understanding the importance of how the structure of the problems led to interactions between scaffolder and learner. In the following excerpts, pattern blocks were used to assist teacher-learners with their understanding of fraction concepts. Pattern blocks were chosen because of their geometric orientation in which fraction concepts would not be separated from their historical origin in measurement (Davydov, 1991). This approach was implemented to deepen the understanding of fraction concepts and to assist the teacher-learners' metacognitive thinking. Our investigation of the conceptual structure of fractions took into account this position as shown in the following:

Facilitator: Let's try another one: Use pattern blocks to rename $\frac{16}{12}$ as a mixed number.

Omar: It's one and two-thirds.

Tammie: No, it's one and four-sixteenths.

Omar: No, I said it's one and two-thirds!

Tammie: I was using sixteenths. So for the record, I'll change my answer to twelfths.

Facilitator: You have one and four-twelfths, if you use twelfths... So, what is the answer?

Omar: Two-thirds.

Tammie: Sixth-twelfths.

Facilitator: Omar, why are you saying two-thirds and Tammie why do you say sixth-twelfths? Think about it. Look at the blocks. Do you need to identify what is the whole? Think about which blocks represent sixteen-twelfths.

Omar: It's two-sixths.

Facilitator: And if you simplify it, it will be?

Omar: One-third.

Facilitator: Very good, and ...?

Omar: It's one and one-third.

Tammie: I still don't get it!

Facilitator: What you have been doing is simplifying a fraction. That's the beauty of using the pattern blocks; they really help you understand how to simplify fractions … a fraction is written in lowest terms when it has the lowest denominator possible … Here we have one, our whole … and what we have here is $\frac{1}{4}$, then $\frac{2}{4}$, which is equivalent to $\frac{6}{12}$. Both of these, $\frac{2}{4}$ and $\frac{6}{12}$, represent $\frac{1}{2}$, which is in the simplest form for $\frac{2}{4}$ and $\frac{6}{12}$.

As the facilitator provided this information to Tammie and Omar, they nodded their heads as well as pointed to the various blocks to communicate their understanding of what is being conveyed. The facilitator continued with an explanation of instruction for when using patterns blocks with their students. The facilitator elaborated, When you have $\frac{2}{4}$ and you have $\frac{6}{12}$, you need to find the fewest blocks possible to represent that fraction. Which block would you use to represent these two equivalent fractions? Next, you need to write an equivalent fraction in its simplest form … $\frac{2}{4}$ in its simplest form is $\frac{1}{2}$, and $\frac{6}{12}$ in its simplest form is $\frac{1}{2}$. In this way, students can see the abstract representation as well as a concrete representation of the fractions. Let's try another one. Which is simpler: $\frac{1}{6}$ or $\frac{2}{12}$?

Omar: One-sixth.

In the excerpt above, note that Tammie not only lacks the abstract understanding of how to rename an improper fraction as a mixed number using an algorithm, but she also lacks the conceptual understanding of what this algorithm represents. Omar initially provides and models the wrong answer, but he eventually comes to both a concrete and abstract understanding of simplifying mixed numbers. The fact that both teacher-learners initially lacked the ability to simplify mixed numbers is alarming, because these individuals have been teaching their students to perform an operation that they do not fully understand themselves. How can we expect *all* students to understand concepts and skills in the abstract when they may not have experienced learning via the trajectory path: concrete, pictorial and abstract? The most notable way these teacher-learners came to a conceptual understanding of mixed numbers was as a direct result of the facilitator clarifying misunderstandings, reformulating approaches, suggesting ways to connect the blocks with the abstract representation, and summarizing and drawing the steps taken along the way. The facilitator's actions and dialogue scaffolded the teachers to a higher level of cognitive understanding (Murray & McPherson, 2006).

A dynamic cognitive assessment model requires the scaffolder to actively monitor and scaffold learning and understanding of mathematics content (Albert, 2002). According to Vygotsky (1978), assessing a learner in the zone of proximal development provides a better prediction of that learner's future performance than results obtained from a conventional test that does not consider a learner's

ZPD. With dynamic cognitive assessment, learners' knowledge and under-standing grow and change as they make meaning of contextually mediated mathematics practices and activities. This aspect is often valued when discussing and monitoring children's learning, but is seldom mentioned when discussing or describing teacher-learners' thinking and understanding. Research shows that "adults learn best when they are interested, feel connected with the topic, feel supported through the learning process, and are able to implement what they've learned" (Nevills, 2003, p. 23). Garet et al. (2001) reported that regarding profes-sional development, the strongest factor related to reported changes in teacher behavior is focused on content knowledge – addressing specific strategies for specific content areas and opportunities for active learning. It follows that an emphasis on social context and collaboration is necessary for meaningful mathematics professional development activities.

Teacher-Learner Talk and the More Knowledgeable Other

As the teachers-learners continued in their development of a conceptual understanding of fractions, they required fewer pedagogical interventions by the facilitator. For example, we asked them to use pattern blocks to find the difference of $\frac{3}{4} - \frac{2}{3}$. First, teacher-learners were to represent the problem concretely with the pattern blocks. Second, they drew an illustration to show how the problem was solved, which included the paper-and-pencil algorithm. The next two excerpts provide some sense of how the teacher-learners approached this problem and the role of the more knowledgeable learner.

Alice: Would you agree that's $\frac{3}{4}$ (*pointing to the red pattern blocks as illustrated in Fig. 4.3*)?

Thomas: Yes … so then minus $\frac{2}{3}$ (*uses $\frac{1}{6}$ pattern blocks to display fraction*). So my thought here, Alice, is that we have to get it all into blue.

Alice: Why do we get it into blue?

Thomas: Well, this is a 1/6, isn't it? And 1, 2, 3, 4… 4/6 is equal to $\frac{2}{3}$, isn't it?

Alice: Yes.

Thomas: Well, what I figure is that we can get this into blues, as well. Do you have any blues I can borrow?

Alice: I would think, looking at the problem, that twelfths should be used.

Thomas: We'll see. You may be absolutely right. You know why, because we can-not do it with this green block.

Alice: So it has to be twelfths?

Thomas: Umm, I don't know. I just made this right here, right? And so how many are we taking away from this?

Alice: You're taking $\frac{2}{3}$ away from it.

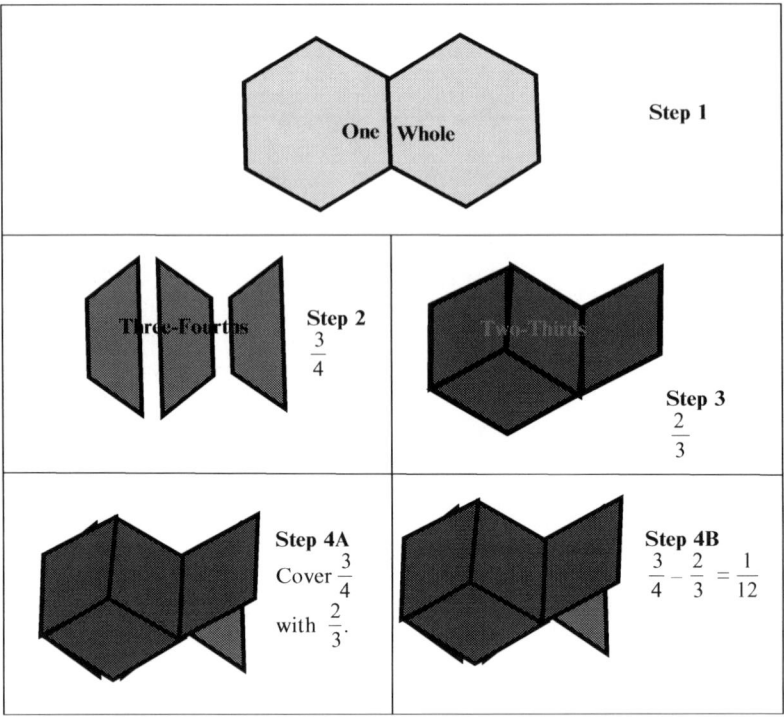

Fig. 4.3 Solution to $\dfrac{3}{4} - \dfrac{2}{3}$ without finding common denominators

Thomas: So we're taking all the blues away, aren't we?
Alice: Right.
Thomas: Oh yell, nice. Does that work for you?
Alice: Yeah, that works. So you did have to find common denominators.
Thomas: Umm, no … I don't think so. But, do you want to do all of these in green? You could certainly do these in green.
Alice: That's the way [the facilitator] has taught us to do it. Well, if that's the way you want your students to do it.
Thomas: To put them all into greens? (*This aspect is illustrated in Fig.* 4.4.)

Thomas is correct in his statement that one can employ pattern blocks to solve the problem without using common denominators (See Fig. 4.3); however, if one is trying to solve this problem using the arithmetic algorithm, one must find a common denominator of 12 (Fig. 4.4). Thomas noticed that he could view three-fourths as three red blocks. He also saw that two-thirds is equivalent to four-sixths, so he used four blue blocks to represent two-thirds. To evaluate the expression $\dfrac{3}{4} - \dfrac{2}{3}$, Thomas covered the three red blocks with the four blue blocks, and noticed that one corner is left uncovered. He saw that this one corner is the difference $\dfrac{3}{4} - \dfrac{2}{3}$, which can be

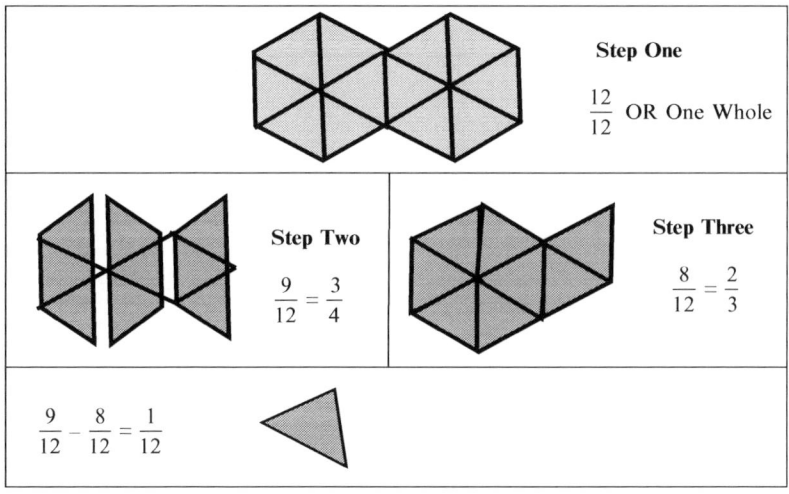

Fig. 4.4 Solution to $\dfrac{3}{4} - \dfrac{2}{3}$ with common denominators

covered by a green block that represents $\dfrac{1}{12}$. Without finding $\dfrac{3}{4}$ and $\dfrac{2}{3}$ to their

equivalent fractions with a denominator of 12, Thomas was able to use the fraction

blocks and see that $\dfrac{3}{4} - \dfrac{2}{3} = \dfrac{1}{12}$. This method gives the teacher-learners a visual

understanding of the processes, but does not give them a concrete representation of the algorithm of which they are familiar. Figure 4.3 presents a systematic illustration of Thomas' approach to the problem.

Further analysis revealed that in the previous excerpts, the facilitator was most involved in the initial stages of instruction. As the teacher-learners became familiar with the activity, they took on the task of scaffolding each other. When the zone of proximal development is created, attention is placed on the interactions between the facilitator, or a more knowledgeable learner, and the less knowledgeable learner. Bruner (1987) writes,

> Once the concept is explicated in dialogue, the learner is enabled to reflect on the dialogue, to use its distinctions and connections to reformulate his own thought. Thought[s], then are both an individual achievement and a social one. … There is another outcome that results from such 'assisted learning,' that bears upon consciousness and volition. For when one climbs to higher conceptual ground—as in going from arithmetic to algebra with the aid of a teacher—one achieves conscious control of the knowledge (p. 4).

The dialogue between Thomas and Alice modeled Bruner's notion of assisted learning. In order for this to happen, Thomas and Alice and later Erin, as illustrated in the next excerpt, assumed four roles[2] at different times: *listener, observer, speaker,*

---

[2] The four roles identified in this chapter were also evidence in prior research on children collaborative learning (Albert & McKee, 2001).

and *analyzer*. Each teacher-learner was able to be an active *listener*, paying attention to what the others were saying. The teacher-learners were *observers*, especially when they watched others manipulating the pattern blocks. When Thomas was actively manipulating, Alice and Erin were observing, asking for clarification, and demonstrating an understanding of what was being shown to them. When the time was right, Alice and Erin became the *speakers*, offering justifications for their decisions through persuasive and well thought-out questions and suggestions. Finally, it was crucial that the teacher-learners *analyzed* as much as possible what they had heard, observed, and talked about when they engaged in dialogue and the manipulation of the blocks to complete the problem. The findings stated here are similar to Bruner's idea that "Thought[s] … are both an individual achievement and a social one." Therefore, the teacher-learner's thoughts had to be considered as to the fitness to the individual and for the intentions of the problem. Each teacher-learner must play each of these roles; when one teacher-learner is speaking, the others must be listening, observing, and analyzing. These roles clearly offer the teacher-learner the opportunity to shift in and out as the *more knowledgeable other*, when it is appropriate according to individual understandings of the task.

Erin further models these aspects in her stance about the role of common denominators in solving the problem. Erin asked Thomas:

Do you want [your students] to use common denominators, to understand common denominators, Thomas?

Thomas:	Yes, I would like them to understand that. But I'm just looking at this and I'm making this incredible visual connection; I'm seeing this here needs to be taken away and I've got it here and it just seems so attractive to me, almost irresistibly attractive, to just remove that and then I look and I think, this is $\dfrac{1}{16}$. And I get the answer that way. It seems to me that converting all of these to green just to take them away would be …
Erin:	Or to pile them on top?
Alice:	I don't think that students would convert them all to green, though, I do agree with that.
Thomas:	You think they'd just go with blue? Well, we could model it both ways since we're here. We could model it in sixteenths … See, I'm tempted to just look up at the board and find out what the common denominator should be just by looking at it and then work backwards, but I'm refusing to do that because I want to see if I can do it this way. I do have a strong preference to keeping the blue.
Alice:	I think that [students] would do it this way with the blue.
Thomas:	I think they would go with the blue, too, just because it's easier to see. Let me step back a moment and think of what's going on here. (*Repeats original problem.*) I want the model for one whole in front of me.
Alice:	Three-fourths is the red.
Erin:	And the three-fourths? Is that intuitive, though, that is $\dfrac{2}{3}$ (*referring to two blue blocks as illustrated in Fig.* 4.3)?

Thomas:	And two-thirds is definitely the blue, because I know that each one of these blues is $\frac{1}{3}$. Well, I look at this and I just started thinking, I just think of it in terms of one. I know that this is $\frac{2}{3}$, in terms of one of these blocks.
Erin:	But the whole thing is one.
Thomas:	Right, when I'm looking at $\frac{2}{3}$, I say, if we're dealing with 1 block, I can see that very easily. This is $\frac{2}{3}$. So, because we're dealing with 2 blocks, I think, "Double whatever I get." So I thought we need 2 – no, we need 4 because we're doubling it ... so I just put 4 of them, so I made it any pattern I wanted to with 4 (*See Fig.* 4.3).
Alice:	You know what you could do, is you could take this, and put it on there, and see that the triangle is all that's left on it.
Thomas:	Oh yeah! I like it!
Alice:	That way, I think, would be more intuitive for them.
Erin:	And your particular preference is piling up, or stacking. How can you concretely represent the abstract?

In the above excerpt, Erin acted as *speaker and analyzer* when she asked, "Do you want [students] to use common denominators, to understand common denominators, Thomas?" She also was simultaneously an *observer*, because she was scrutinizing the work completed by Thomas. Thomas and Alice were *listeners*, actively nodding to show that they were paying attention to Erin's question. Thomas also was the *analyzer* when he stated that he wanted his students to understand common denominators, providing an explanation of why. Alice became the *speaker* when she provided an explanation to Erin's question. Alice was also demonstrating proof of her own ability to *analyze* the situation; she obviously thought about Erin's questions as well as the clarification statement made by Thomas. At the same time, Alice was acting as the more knowledgeable other, sharing her knowledge with Thomas and Erin and scaffolding him into recalling information that had previously been explained. "But the whole thing is one," she replied to Thomas. To assist the teacher-learners in clarifying their thinking the facilitator intervened:

Did you explain your solution? Can you show the paper-and pencil-algorithm? Oh, I see you have found the solution and you are contemplating multiple approaches to the solution.

Alice:	My idea is that you would want [students] to use common denominators, but we were just discussing that they would not do that, and we were trying to figure out what exactly they would be doing.
Facilitator:	Okay, would you use common denominators with the paper-and-pencil algorithm?
Alice:	I would.
Thomas:	Maybe, I would.

Figure 4.4 shows that the teacher-learners would have arrived at the same answer had they taken Erin's advice to "concretely represent the abstract [algorithm]." While the method illustrated in Fig. 4.3 does lead to the correct response, it does not give a concrete representation of the arithmetic paper-and-pencil process students use to subtract fractions with unlike denominators. This is simply because it does not require the use of common denominators. Erin's method provides a concrete representation of what it *looked like* to find a common denominator. The teacher-learners would have to change $\frac{3}{4}$ to $\frac{9}{12}$ and $\frac{2}{3}$ to $\frac{8}{12}$. They could have used green blocks to represent these fractions, and would have noticed that $\frac{9}{12} - \frac{8}{12}$ is equal to $\frac{1}{12}$, which is equivalent to $\frac{3}{4} - \frac{2}{3}$. While both methods are correct, Erin's method provides a more comprehensive representation of how one subtracts fractions. Because her representation includes the notion of finding common denominators, it more closely resembles the actual algorithmic process of subtracting fractions. However, this notion is similar to a conclusion drawn by Thornton (1995). Thornton writes that:

> The discovery of variation in skill from one context to another has radically changed our understanding of what is involved in problem solving. Instead of being driven by abstract skills like logic, problem solving draws deeply on knowledge of the particular concrete detail of the task in hand. What you know about a task determines how you plan to tackle the problem, what strategies you consider, how you interpret feedback (p. 120).

The teacher-learners' skills varied when they used concrete materials as they worked in small groups to illustrate their knowledge and understanding of the algorithm. Their application of the abstract representation of the algorithm when solving it via paper-and-pencil depended on an abstract general process they were more familiar with from past learning experiences.

Discussion of Findings

As noted earlier in this chapter, this inquiry was predicated on the notion that scaffolded instruction situated in professional development seminars actively engages teachers as learners and scaffolders in which the critical element is the use of language for mediation (Vygotsky, 1978; Wertsch, 1980). Language affords the means for interactions between the scaffolder and learner allowing for individual as well as co-construction of knowledge. According to Bauersfeld (1995), although there is an intrinsic interdependence between scaffolding and learning, it is imperative that teachers contribute to the contexts, developing a view that their learning and understanding of the content can happen with the support of knowledgeable others. In the context of the professional development seminars, this feature may in part explain why there was improvement from the pretest to posttest scores, with statistically significant improvement on the group open-response item of the *Content Knowledge Test*.

A further explanation that provides insight regarding why the teachers showed improvement in their content knowledge may be that the structure of the seminars functioned to augment rather than restrict teacher learning. Participants learned how to communicate and scaffold in various ways as the more knowledgeable other. Findings indicated that an essential element to being the knowledgeable other rests on the teacher-learners taking on four roles: *speaker, listener, observer,* and *analyzer*. These roles clearly offered each teacher-learner the opportunity to shift in and out as the more *knowledgeable other* when it was appropriate to scaffolding understandings of the mathematical content. The scaffolding that occurred among the teacher-learners helped them in focusing, framing, and enhancing their learning of the mathematical content to make tangible links to their own teaching practices. Interaction between the teacher-learners provided a context for the tool of language to function as a bridge between the mathematical task and their prior knowledge, thereby supporting the learning process.

Connecting Learning to the Field of Social Interaction

Acknowledging the role of language in the learning process for the teacher-learners exposes a critical issue. If we only consider the results of the *Content Knowledge Tests*, it would not be difficult to conclude that by the end of the study the teacher-learners possessed the content knowledge they needed to teach at the middle school level. This finding and subsequent conclusion did not reveal the challenges that some of the teacher-learners experienced in understanding the content. This was most evident during the number and operations seminars. It was the *Field of Social Interaction* that served as a useful context for exchange of mathematical thoughts, which assisted the teacher-learners in overcoming the lack of experience with unfamiliar strategies or tools (i.e., concrete material) and made it easier for them to avoid surface level explanations not connected to mathematical problems.

It is important to note that although the facilitator was an essential part of the *Field of Social Interaction,* the facilitator did not play a dominant role in the learning process when the teacher-learners were interacting in small collaborative groups or working in pairs. The facilitator often set the problem or task in motion and then stepped back into the role of observer, intervening only when the teacher-learners became completely lost or involved in severe inconsistencies about the problem at hand. The facilitator maintained the momentum of the seminars by holding the teacher-learners accountable for their own learning through questions and hints. This position by the facilitator was intentional because successful scaffolding, along with holding the teacher-learners accountable, necessitated deliberate actions by the facilitator; additionally, it required a context where learners actively engaged in activities that carry through to successful mathematical understanding. It was within this context that zones of proximal development were created in the course of dialogue, instruction, and interaction. As stated, the facilitator was most involved in the initial stages of the problem. As the teacher-learners became familiar with the problem, they began to take on the task of scaffolding each other, relying less on the facilitator. When zones of proximal development were created during the seminars, consideration was given to the interaction between the facilitator, or a more knowledgeable learner, and the less knowledgeable learner. This was a key component because when considering the learning and understanding of these teacher-learners, the zone of proximal development provided an augmented sign for "predicting or understanding future intellectual development" (Chaiklin, 2003, p. 56).

These findings are consistent with the work of Bruner (1987) regarding the role of the teacher in assisted learning situations. Specifically, Bruner writes that academic performance is both an individual and social achievement produced by assisted learning that is the result of conscious and intentional control of knowledge. Furthermore, it was not surprising that by the conclusion of the study, there was an increase in the percentage of teacher-learners stating that they preferred learning new material and engaging in problem solving tasks collaboratively. The teacher-learners consistently elaborated on the postsurvey that the most significant benefits of collaborative inquiry were understanding the importance of multiple perspectives, developing knowledge and understanding of student learning, creating potential to improve pedagogical practices, and creating potential to improve knowledge construction of content. Thus an insightful observation is that their participation during the seminars assisted them in appreciating that "One builds a learning community ... engendering commitment in individual not by manipulating control" (Prawat, 1996, p. 101). They realized that their experiences in the professional development seminars could translate to teaching experiences in their classrooms.

The Importance of Language and Action

The deliberate and intentional nature of scaffolded instruction is not only consistent with the work of Bruner, but it also mirrors the work of Vygotsky (1978, 1994) and Bakhtin (1984) regarding the role of language in learning and development. The process of learning is both individual and sociocultural, the use of language is crucial in helping both the individual and the collective make sense of experiences. For these teacher-learners, language scaffolded or mediated their *actions* as they participated in collaborative practices. Action is emphasized here because Vygotsky and Luria (1994) strongly suggested that the importance of not ignoring an action is continuously linked to speech. If we exclude the inflated feature of language ("the word"), then what surfaces is an "underestimation of volitional [preference or desire] action, action in its highest forms, that is, action tied to the word" (Vygotsky & Luria, p. 169), which plays a major role in the development of higher cognitive functions (i.e., evaluating the reasonableness of a solution).

When those of us who plan professional development experiences for teachers listen to them and seriously consider their dialogue and action, then scaffolded instruction may provide fresh perspectives on how to help them participate more fully in their learning, to become more effective mathematics teachers, and to become the positive inheritors of the future for mathematical learning of our students. A pertinent argument is that the teacher's actions and speech within professional development contexts may be a better source for understanding the relationship that exists between mathematics teaching and learning. Bakhtin's (1984) notion of the influence of speech on word meaning has some bearing on this argument. Bakhtin proposed that the channel of communication, and for this study, the professional development seminars, "must not be separated from the realm of discourse, that is, from language as a concrete integral phenomenon. Language is only in the dialogic [or conversation] interaction of those who make use of it" (Bakhtin, p. 183). Therefore, actual dialogic interaction must be grounded in the relationship that permeates social discourse with people and must not be separated from it.

As Vygotsky (1978) projected, "all the higher functions originate as actual relations between [people]" (p. 57). The major argument rests on the assumption that development cannot be separated from social contexts or from language, oral or written. Therefore, the starting point for these teacher-learners was geared toward having them engage in talk, task, and content with which they were familiar – the traditional nature of mathematics teaching and learning. However, the intention was to move their understanding and thinking forward to a higher intellectual level, not just about the content's material but also about how to teach the content. Language, Vygotsky and Luria (1994) insisted, along the way become:

> Intellectualized and developed on the basis of *action*, lift this action to a supreme level ... If *at the beginning* of development there stands the act, independent of the word, then at the end of it there stands the word which becomes the act, the work which makes [individual's] action free (p. 170, emphasis in original).

During the seminars, when the facilitator applied deliberate pedagogical practices for learning and understanding mathematics content, understanding took on personal and intellectual qualities for the teacher-learners. The facilitator, as the knowledgeable other scaffolded learning and understanding, gradually allowing the teacher-learners to monitor and regulate their own thinking about and learning of the material, decide the appropriateness of different strategies, and successfully complete given problems, independently or collectively.

Concluding Comments

In this chapter, we have argued that language in scaffolded instruction assists teacher-learners by offering a foundation from which they can build their understanding of mathematical content. This was exemplified through excerpts of teacher-learners' interactions and talk with each other as they engaged in learning tasks during the scaffolding process. Results of the *content knowledge tests* indicated that the teachers possessed the mathematics content knowledge needed to teach at the middle school level. However, this does not mean that they have the pedagogical knowledge or skills for teaching mathematics that goes beyond traditional methods.

What the *content knowledge tests* did not reveal was the importance of providing professional development experiences that include teaching mathematics concepts with concrete material, fostering effective collaborative learning, and recognizing that successful scaffolded instruction requires that teachers know the content thoroughly. Thus, they must come to understand that there is not an either/or proposition; understanding the content must come together with pedagogy for teaching that content. It is essential that teachers create collaborative groups that encourage participation of all learners to limit barriers in developing understanding of content knowledge. They must also recognize that intentional and deliberate scaffolded instruction can provide opportunities for all learners to be the knowledgeable other.

Throughout this chapter, we emphasize that scaffolded instruction and collaborative discourse are necessary elements for meaningful mathematics professional development activities. The language used in these contexts fosters meaningful professional development because of the exchange of ideas and actions for which it allowed. A potential implication is that language and action in professional development activities may assist teachers in building a community of practice to communicate, reason, and talk with precision about mathematics concepts and skills; to develop an everyday practice of thinking metacognitively; and to deepen content and pedagogical knowledge used in the practice of teaching and learning, such as choosing and using various representations of mathematics to further students' learning. Thus, it is worth noting that when teachers were provided with professional development experiences that focused on collaborative group work and mathematics content, their preference for working individually shifted to preferring to work collaboratively on problem solving activities. The professional development seminars engendered a learning community in which the teachers had the opportunity to enhance strategies and techniques, which actualized greater intellectual understanding about themselves as learners and scaffolders. The teachers came to understand that they are the constructors of their own intellectual development, actively engaging in the process of exploring and making mathematical assumptions as well as interpreting and organizing information for use and knowledge construction in their classrooms.

It is clearly impossible to incorporate all the issues and perspectives surrounding this study. Consequently, the presentation of findings was highly selective; we acknowledge that many issues and meanings of language and action, content learning in social contexts, and scaffolded instruction and collaborative discourse are not covered here. It is recognized that the interpretations and discussion upon which the findings presented in this study are based on a small sampling of participants. This study's findings, however, do suggest the importance of creating learning communities that promote and connect teacher learning and the development of content and pedagogical knowledge. This research provides a fresh perspective on the role of learning and understanding mathematics content within a collaborative context in which teachers' metacognitive processes evolve and influence their role as teachers of mathematics. A promising contribution of this study is the representation of teacher learning processes vis-à-vis concrete models and dialogue, thereby enabling a process of self-understanding and self-reflection that may transform teachers' conceptions of the way students are learning; this may ultimately produce a different image of themselves as teachers of mathematics.

References

Albert, L. R. (2000). Outside in, inside out: Seventh grade students' mathematical thought processes. *Educational Studies in Mathematics, 41*, 109–142.

Albert, L. R. (2002). Bridging the achievement gap in mathematics: Sociocultural historic theory and dynamic cognitive assessment. *Journal of Thought, 37*, 65–82.

Albert, L. R., Mayotte, G., & Phelan, C. (2004). The talk of scaffolding: Communication that brings adult learners to deeper levels of mathematical understanding. In D. E. McDougal & J. A. Ross (Eds.), *Proceeding of the twenty-sixth annual meeting of the American Chapter of the International Group for the Psychology of Mathematics Education* (pp. 1137–1138). Toronto, Canada: OISE/UT.

Albert, L. R., & McKee, K. (2001). In their own words: Achieving intersubjectivity through complex instruction. In V. Spiridonov, I. Bezmenova, O. Kuoleva, E. Shurukht, & S. Lifanova (Eds.), *The summer psychology conference 2000, the zone of proximal development* (pp. 6–23). Moscow: Institute of Psychology of the Russian State University for the Humanities.

Andrade, A. D. (2009). Interpretive research aiming at theory building: Adopting and adapting the case study design. *The Qualitative Report, 14*, 42–60.

Bakhtin, M. (1984). *Problems of Dostoevsky's poetics*. Minneapolis, MN: University of Minnesota Press.

Bauersfeld, H. (1995). Language games' in the mathematics classroom: Their function and their effects. In P. Cobb & H. Bauersfeld (Eds.), *The emergence of mathematical meaning: Interaction in classroom cultures* (pp. 271–289). Mahwah, NJ: Lawrence Erlbaum Associates.

Bruner, J. (1987). Prologue. In *L. Vygotsky, the collected works of L. S. Vygotsky* (M. Cole, S. Scribner, V. John-Steiner, & E. Souberman, Trans.). Cambridge, MA: Harvard University Press.

Chaiklin, S. (2003). The zone of proximal development in Vygotsky's analysis of learning and instruction. In A. Kozulin, B. Gindis, V. Ageyev, & S. Miller (Eds.), *Vygotsky's educational theory in cultural context* (pp. 39–64). New York: Cambridge University Press.

Cobb, P. (1994). Where is the mind? Constructivist and sociocultural perspectives on mathematical development. *Educational Researcher, 23*(7), 13–20.

Cohen, E. (1994). *Designing group work: Strategies for the heterogeneous classroom*. New York: Teachers College Press.

Cohen, D. K., & Hill, H. (2001). *Learning policy: When state education reform works*. New Haven, CT: Yale University Press.

Creswell, J. W. (2003). *Research design: Qualitative, quantitative, quantitative, and mixed approaches*. Thousand Oaks, CA: Sage.

Creswell, J. W., & Plano Clark, V. L. (2006). *Designing and conducting mixed methods research*. Thousand Oaks, CA: Sage.

Creswell, J. W., Plano Clark, V. L., Guttmann, M. L., & Hanson, E. E. (2003). Advanced nixed methods research design. In A. Tashakkori & C. Teddlie (Eds.), *Handbook of mixed methods in social and behavioral research* (pp. 209–240). Thousand Oaks, CA: Sage.

Davydov, V. V. (1990). *Types of generalization in instruction*. Reston, VA: National Council of Teachers of Mathematics.

Davydov, V. V. (1991). On the objective origin of the concept of fractions. *Focus on Learning Problem in Mathematics, 13*(1), 13–64.

Davydov, V. V. (1995). The influence of L. S. Vygotsky on education theory, research, and practice. *Educational Researcher, 24*, 12–21.

Davydov, V. V. (1998). The concept of developmental teaching. *Journal of Russian and East European Psychology, 36*(4), 11–36.

Doolittle, P. (1997). Vygotsky's zone of proximal development as a theoretical foundation for cooperative learning. *Journal on Excellence in College Teaching, 8*(1), 83–103.

Gall, M. D., Gall, J. P., & Borg, W. R. (2010). *Applying educational research.* Boston: Pearson.

Garet, M. S., Porter, A. C., Desimore, L., Birman, B. F., & Yoon, K. S. (2001). What makes professional development effective? Results from a national sample of teachers. *American Educational Research Journal, 38*(4), 915–945.

Gay, L. R., & Airasian, P. (1996). *Educational research: Competencies for analysis and application.* Upper Saddle River, NJ: Merrill.

Goos, M. (1999). Scaffolds for learning: A sociocultural approach to reforming mathematics teaching and teacher education. *Mathematics Teacher Education and Development, 1,* 4–21.

Goos, M. (2004). Learning mathematics in a classroom community of inquiry. *Journal for Research in Mathematics Education, 35,* 258–291.

Goos, M. (2005). A sociocultural analysis of the development of pre-service beginning teachers' pedagogical identities as users of technology. *Journal of Mathematics Teacher Education, 8*(1), 35–59.

Greene, J. C., Caracelli, V. J., & Graham, W. F. (1989). Toward a conceptual framework for mixed-method evaluation designs. *Educational Evaluation and Policy Analysis, 11*(3), 255–274.

Hill, H. C. (2004). Professional development standards and practices in elementary school mathematics. *The Elementary School Journal, 104,* 345–363.

Hill, H. C., & Ball, D. (2004). Learning mathematics for teaching results from California's mathematics professional development institutes. *Journal for Research in Mathematics Education, 35,* 330–351.

Hogan, K., & Pressley, M. (1997). Scaffolding scientific competencies within classroom communities of inquiry. In K. Hogan & M. Pressley (Eds.), *Scaffolding student learning: Instructional approaches and issues* (pp. 74–107). Cambridge, MA: Brookline Books.

Janesick, V. J. (1994). The dance of qualitative research design: Metaphor, methodology, and meaning. In N. K. Denzin & Y. S. Lincoln (Eds.), *Handbook of qualitative research* (pp. 209–219). Thousand Oaks, CA: Sage Publications, Inc.

Jennings, C., & Di, X. (1996). Collaborative learning and thinking: The Vygotskian approach. In L. Dixon-Krauss (Ed.), *Vygotsky in the classroom: Mediated literacy instruction and assessment* (pp. 77–91). New York: Longman Publishers.

Johnson, R. B., & Onwuegbuzie, A. J. (2004). Mixed methods research: A research paradigm whose time has come. *Educational Researcher, 33*(7), 14–26.

Kozulin, A. (1998). *Psychological tools: A sociocultural approach to education.* Cambridge, MA: Harvard University Press.

Kumpulainen, K., & Mutanen, M. (2000). Mapping the dynamics of peer group interaction: A method of analysis of socially shared learning processes. In H. Cowie & G. van der Aalsvoort (Eds.), *Social interaction in learning and instruction: The meaning of discourse for the construction of knowledge* (pp. 144–160). Amsterdam: Pergamon.

Larkin, M. J. (2001). Providing support for student independence through scaffolded instruction. *Teaching Exceptional Children, 34,* 30–34.

Miles, M. B., & Huberman, A. M. (1994). *Qualitative data analysis.* Thousand Oaks, CA: SAGE Publications.

Murray, D. E., & McPherson. (2006). Scaffolding instruction for reading the web. *Language Teaching Research, 10,* 131–156.

National Research Council. (2002). *Scientific research in education.* Washington, DC: National Academy of Sciences.

Nevills, P. (2003, Winter). Cruising the cerebral superhighway. *Journal of Staff Development, 24*(1), 20–23.

Osana, H., & Folger, T. (2000). *Negotiated meaning in small group conversation: Talk in a schools for thought classroom*. Paper presented at the 2000 annual meeting of the American Educational Research Association, New Orleans, LA.

Palinscar, A. (1986). The role of dialogue in providing scaffolded instruction. *Educational Psychologist, 21*(1–2), 73–98.

Palinscar, A., & Brown, A. (1988). Teaching and practicing thinking skills to promote comprehension in the context of group problem solving. *RASE, 9*(1), 33–39.

Prawat, R. S. (1996). Learning community, commitment, and school reform. *Journal of Curriculum Studies, 28*(1), 91–110.

Roehler, L., & Cantlon, D. (1997). Scaffolding: A powerful tool in social constructivist classrooms. In K. Hogan & M. Pressley (Eds.), *Scaffolding student learning: Instructional approaches and issues* (pp. 6–42). Cambridge, MA: Brookline Books.

Rojas-Drummond, S. (2000). Guided participation, discourse and the construction of knowledge in Mexican classrooms. In H. Cowie & G. van der Aalsvoort (Eds.), *Social interaction in learning and instruction: The meaning of discourse for the construction of knowledge* (pp. 193–213). Amsterdam: Pergamon.

Rosenshine, B., & Meister, C. (1992, April). The use of scaffolds for teaching higher-level cognitive strategies. *Educational Leadership, 49*, 26–33.

Thorne, S., Kirkham, S. R., & O'Flynn-Magee, K. (2004). The analytic challenge in interpretive description. *International Journal of Qualitative Methods, 3*(1), 1–11.

Thornton, S. (1995). *Children solving problems*. Cambridge, MA: Harvard University Press.

Vygotsky, L. S. (1978). *Mind in society: The development of higher psychological processes*. Cambridge, MA: Harvard University Press.

Vygotsky, L. S. (1994). The problem of the environment. In R. Van Der Veer & J. Valsiner (Eds.), *The Vygotsky reader* (pp. 338–354). Cambridge, MA: Blackwell.

Vygotsky, L. S., & Luria, A. (1994). Tool and symbol in child development. In R. Van Der Veer & J. Valsiner (Eds.), *The Vygotsky reader* (pp. 99–174). Cambridge, MA: Blackwell.

Wasser, J., & Bresler, L. (1996). Working in the interpretive zone: Conceptualizing collaboration in qualitative research teams. *Educational Researcher, 25*, 5–15.

Wegerif, R., & Mercer, N. (2000). Language for thinking: A study of children solving reasoning test problems together. In H. Cowie & G. van der Aalsvoort (Eds.), *Social interaction in learning and instruction: The meaning of discourse for the construction of knowledge* (pp. 179–192). Amsterdam: Pergamon.

Wells, G. (1999). *Dialogic inquiry: Towards a sociocultural practice and theory of education*. New York: Cambridge University Press.

Wells, G. (2000). Dialogic inquiry in education: Building on the legacy of Vygotsky. In C. D. Lee & P. Smagorinsky (Eds.), *Vygotskian perspectives on literacy research* (pp. 51–85). New York: Cambridge University Press.

Wertsch, J. V. (1979). From social interaction to higher psychological processes: A classification and application of Vygotsky's theory. *Human Development, 22*, 1–22.

Wertsch, J. V. (1980). The significance of dialogue in Vygotsky's account of social, egocentric, and inner speech. *Contemporary Educational Psychology, 5*, 150–162.

Wood, D., Bruner, J., & Ross, G. (1976). The role of tutoring in problem solving. *Journal of Child Psychology and Psychiatry, 17*, 89–100.

Chapter 5
Closing Thoughts and Implications

As illustrated in the previous chapters, Vygotskian theory serves as a scintillating discussion for understanding the value of framing mathematics education pedagogy and research in sociocultural contexts that permeate our daily actualities, our learning of concepts, our dynamic engagement in activities, and our use of language. For example, teacher-generated drawings as a phenomenon unmask the values and concepts of a coherent mathematics system, a system that does not easily allow its learners to see the world with a different point of view. The drawings may serve as metaphors, representing learning experiences that are "characterization[s] of a phenomenon in familiar terms that [are] graphic, visible, and physical in our scale of the world" (Dickmeyer, 1989, p. 151). What's more, the studies presented provide a point of view regarding the significance of understanding the origin or history of teachers' mathematical learning. This initial understanding of the role of history in learning extends to Vygotsky's idea about how sociocultural contexts are essential for learning and development, emphasizing a recursive process for intellectual construction or transformation of knowledge. Our closing thoughts bring together major assumptions about Vygotskian theory and mathematical learning discussed in the previous chapters. We focus primarily on tools and signs, intersubjectivity, and the zone of proximal development.

Teacher-Generated Drawings as Tools of Human Development

What we learned from our examination of visual and written data generated by prospective teachers is that the drawings and narratives helped the prospective teachers explore the multiple ways in which these images deepen their understandings of the sociocultural nature of learning. For this reason, they explicitly focused on their own learning as students in various contexts, e.g., an academic classroom or in practice. The goal was to encourage self-reflection among prospective teachers in that they might better understand the preconceptions that they carry into the

L.R. Albert, *Rhetorical Ways of Thinking: Vygotskian Theory
and Mathematical Learning*, DOI 10.1007/978-94-007-4065-5_5,
© Springer Science+Business Media Dordrecht 2012

classroom, and to help them realize that their students are just as heavily thwarted with preconceptions. "While images always maintain some connection to people, places, things or events, their generative potential in a sense gives them a life of their own, so that we not only create images, but are also shaped by them" (Weber & Mitchell, 1996, p. 305). Such awareness, we believe, may raise the quality of teaching that takes place in mathematics classrooms.

We contend that to understand the thinking of prospective teachers, we must become familiar with the social context in which their learning is situated. Images, such as drawings, are created, but shaped by human experience and are valid tools in which to understand how prospective teachers make sense of their work and, in particular, how they understand teaching and learning of mathematics (Bassette, 2008; Finson, 2002). We anticipated that teacher-generated drawings might be used as tools for reflection because they have "strong communicative function" (Weber & Mitchell, 1996, p. 303). Teacher-generated drawings may serve as *communicative tools* for mediating inner thoughts about mathematical teaching and learning, while communicating that meaning to others (Albert, 2000; Katz et al., 2011). "If we want to understand how [mathematics] teachers make sense of their work – to acquire an empathetic understanding from within," argue Efron and Joseph (1994), "then we must explore an artistic form of image that can grasp and reveal the not always definable emotions" (p. 55).

As discussed in detail in Chap. 2, Vygotsky's (1978) notion of the role of *tools and signs* characterized by mathematics learning involves the use of a variety of psychological tools that assist in the thinking process. Vygotsky theorizes that these tools are culturally based and serve as signs that direct us to a deeper understanding of the activity in which the individual is engaged. Thus, Vygotsky's sociocultural historic theory presents a powerful image of human learning. It serves, as an unexpected yet needed metaphor for revealing aspects of the complex nature of teacher-generated drawings in that the drawings serve as a text, calling attention to their fine points, a starting place for understanding and interpreting prospective teachers' conception of themselves as mathematics teachers. Images are constructed and interpreted in attempt to make sense of human experience (Brooks, 2005).

The implication for teacher development in mathematics education is that the drawings and written narratives might help make visible that which might have remained abstract, causing the inner thoughts and images of teaching and learning mathematics become visual representations of abstract ideas and concepts. We argue that they are also illustrative aspects of teacher development in which we come to understand the nature of the learning community and its relative inviolability as the prospective teachers are developing their assumptions about the teaching and learning of mathematics. Their written narratives helped to give a clearer under-standing of what was being expressed in the drawings. Thus the narratives as well as the drawings themselves became a record of how they are developing in their perceptions of mathematical teaching and learning. Their drawing represents an essential step in their development as mathematics teachers.

Achieving Intersubjectivity Through Collaborative Learning

The empirical studies presented in this book are examples of Vygotsky's sociocultural historic theory, which helps explain teaching and learning processes that afford experiences for learners to advance intellectually their mathematical knowledge, skills, and ideas. These teaching and learning processes create collaborative contexts as illustrated in Chap. 4, through which a group of learners begin a task, activity, or discussion with different understandings, but ultimately achieve shared understanding or a *state of intersubjectivity* (Plaskoff, 2003; Rommetveit, 1979). Intersubjectivity results from this interaction as the perspectives of the teacher-learners intertwine, mingle, transform, and coalesce to develop shared meanings. To achieve collaboration and to communicate effectively during joint activity, it was crucial that the teacher-learners worked toward a similar goal (Albert, Bilics, Lerch, & Weaver, 1999, 2000; Bruner, 1996). The interactions that occurred within the professional learning community created the context for socially shared thinking. It is through scaffolded instruction and social interactions that the teacher-learners used *communicative tools* (Albert, 2000) to negotiate meaning as they strived for a shared notion of the mathematics task or problem.

When developing and structuring collaborative contexts for scaffolded instruction or socially shared thinking and learning, it is essential, however, to recognize that teachers need the time and space to talk about the skills and concepts they are teaching in their classrooms, and at the same time they need to become aware of their learning histories. Engaging in such contexts could assist them in understanding that just as they have learning histories that influenced their thinking, understanding, and teaching of mathematics content, so too might be the case for their students; students may have learning histories replete with positive and negative experiences that must be acknowledged. Therefore, sociocultural historic theory as a framework may serve as a catalyst for how to provide experiences that promote dialogue about mathematics teaching and learning. This aspect would include ways to enhance pedagogical content knowledge for teacher-learners by providing a tool through which to appraise their thinking about what mathematical activities worked and did not work, and how they might transform them to make them work for their own profile. Perhaps the most noteworthy suggestion is that sociocultural historic theory as a framework for studying and working with prospective and practicing teachers may influence them in exploring the connection between their learning as teachers and their subsequent teaching of mathematics. We propose that Vygotskian theory renders the structural support new and practicing teachers' need for developing pedagogical plans that make possible effective use of scaffolded instruction and sustained interaction in the classroom. Placing mathematics teachers in experiential learning environments assists them in their professional development as well as helps them to advance their understanding of student learning, providing for a more comprehensive understanding of pedagogical content knowledge.

Adult Learning in the Zone of Proximal Development

In the preceding chapters, we argued that learners bring diverse backgrounds, experiences, knowledge, and understandings about education with them in their own learning situations. Vygotsky's sociocultural historic theory can assist us in understanding how these social and cultural influences affect learning and development. In our discussion, we grappled with key theoretical and practical elements to answer questions such as to what extent are learners' sociocultural experiences relevant to the mathematics learning community? How is learning scaffolded when the sociocultural histories are considered and are used as tools or resources by the learners in mediated activities in learning mathematics? What is the role of the zone of proximal development in scaffolding mathematical understanding and collaborative learning situations? In an attempt to answer these questions and others that emerged during the course of this inquiry, we found that, in part, Vygotsky's (1978, 1986, 1994) sociocultural historic theory provides a useful framework for understanding learning and development of cognitive processes within social contexts and educators' understandings of the various contexts in which learning is situated.

An insightful perspective emerging from our work is the practical illustration of how the zone of proximal development can be broadened to adult learning and development, (e.g., teacher-learners). "An essential feature of learning is that it creates the zone of proximal development; that is, learning awakens a variety of internal developmental processes that are able to operate only when the learner is interacting with people in his environment and in cooperation with his peers" (Vygotsky, 1978, p. 90). The zone of proximal development is the context in which social interaction and *other-assistance* is embedded; the individual learns in collaboration with others (Albert, 2000). Critical to this process is that in adult collaboration, the teacher-learners function as supportive tools and knowledgeable others as they participated in the learning process to construct and co-construct the solutions to problems. What we see here is that the powerful interweaving of individual learning and collective learning is consistent with Vygotsky's (1978, 1981, 1986, 1994) contention that to understand individual development, it is necessary to understand the social contexts in which the individual resides. Individual development involving mediated tools undergoes qualitative changes when it "transitions from a social to the individual function" (1981, p. 159).

What is also evidence from our work is that the more *knowledgeable other* during the initial stages of collaborative learning was an example of teaching by scaffolding, not by *telling* (Albert & Jones, 1997, p. 290; Davydov, 1995, p. 13). The *more knowledge other* must be able to assess the situation, listen to the ideas of the group, and initiate an activity that helps the group move forward in their thinking as learners and problem solvers. To scaffold student learning and recognize that students possess different needs within the zone of proximal development, teachers must know the content thoroughly. They must also be able to create learning situations that encourage participation of all learners. For some learners, it is natural to allow the more knowledgeable other to scaffold their learning, be it a peer or the teacher.

An important aspect of scaffolding learning is being an active listener. That is, learners needed to be diligent about being speakers and listeners. "When ... learners decide to work together, they create a unit that wants to exist as such, a group that must find a point of balance somewhere between silence and a shouting match" (Conle, Louden, & Mildon, 1998, p. 179). We found that by addressing what is and what is not listening may help identify behaviors that lead to active listening and to reflect on those behaviors together. The opportunity to reflect on listening as a group may assist learners in developing higher cognitive functions and increase awareness of the concepts understudy (Albert & Jones).

Another relevant assumption that emerged from our analysis of this work is that prior experiences of working collaboratively on projects influenced interactions. Did the fact that the teacher-learners were adult learners make a difference in collaborative learning situations? That is, children bring less experience to the group situation. "Children have fewer pragmatic life experiences. Learning focuses largely on forming and accumulating basic meanings, values, skills and strategies" (MacKeracher, 1996, p. 19). Children need more focused scaffolded instruction about group process, such as role development and how to function as active and effective group members do. As adult learners, the teacher-learners benefited from the review and reflection processes they already had experienced in many other collaborative situations. An important implication here, then, is that educators at all levels should learn together using a diversity of methodology. Educators of students in the primary grades have insights into a range of concrete introductions to mathematical content knowledge but may lack a complete understanding of the big ideas of mathematics. In contrast, educators of older students and adults who often have a more specialized knowledge of the content may possibly have fixed notions and techniques for teaching that content. Both groups would benefit from professional learning activities to work together to develop their theoretical and practical knowledge about mathematical pedagogy. Our hope is that the theories and work discussed in this book provide concrete examples of how to situate research and various practices and professional activities in sociocultural historic contexts, representing the combination of experiences, collaborative endeavors, and Vygotsky's concept of sociohistorical learning and development. The idea is to keep in mind that mathematical thought occurs outside and beyond the mind – returns to the mind – and then it evolves in another direction based on new and different thoughts rendered through sociocultural experiences.

In conclusion, though Vygotsky is remembered as a psychologist, when examining his theories, it is vital to always keep in mind *Vygotsky the man*, for his goals were more expansive than the mere study of *psychology for psychology's sake*. Rather Vygotsky grounded his theories by focusing on the full spectrum of the human experience in context. This notion is poetically exemplified in his 1931 letter to a student in which he makes the observation: "How much life, warmth, support there is in the quest [towards truth]! And then there is the most important-life itself- the sky, the sun, love, people, suffering. Those are not simply words; it exists. It is real. It is interwoven in life" (van der Veer & Valsiner, 1991, p. 16). He calls his student, and us all, to engage in the study of life in any way possible.

References

Albert, L. R. (2000). Outside-in-inside-out: Seventh-grade students' mathematical thought processes. *Educational Studies in Mathematics, 41*, 109–141.

Albert, L. R., Bilics, A., Lerch, C., & Weaver, B. (1999, April). *The web of intersubjectivity: A Vygotskian framework for adult learning.* Symposium at the American Educational Research Association Annual Meeting, Montreal, Canada.

Albert, L. R., Bilics, A., Lerch, C., & Weaver, B. (2000, April). *Applying the web of intersubjectivity: A Vygotskian framework for adult learning.* Symposium at the American Educational Research Association Annual Meeting, New Orleans, Louisiana.

Albert, L., & Jones, D. (1997). Implementing the science teaching standards through complex instruction: A case study of two teacher-researchers. *School Science and Mathematics, 97*, 383–391.

Bassette, H. J. (2008). Using students' drawings to elicit general and special educators' perceptions of co-teaching. *Teaching and Teacher Education, 24*, 1376–1396.

Brooks, M. (2005). Drawing as a unique mental development tool for young children: Interpersonal and intrapersonal dialogues. *Contemporary Issues in Early Childhood Education, 6*, 80–91.

Bruner, J. (1996). *The culture of education.* Cambridge, MA: Harvard University Press.

Conle, C., Louden, W., & Mildon, D. A. (1998). Tensions and intentions in group inquiry: A joint self-study. In M. L. Hamilton, S. Pinnegar, T. Russell, J. Loughran, & V. LaBoskey (Eds.), *Reconceptualizing teaching practice: Self-study in teacher education* (pp. 178–194). Bristol, PA: Falmer Press.

Davydov, V. V. (1995). The influence of L. S. Vygotsky on education theory, research, and practice. *Educational Researcher, 24*, 12–21.

Dickmeyer, N. (1989). Metaphor, model, and theory in education research. *Teachers College Record, 91*, 151–160.

Efron, S., & Joseph, P. B. (1994). Reflections in a mirror-teacher-generated metaphors from self and others. In P. B. Joseph & G. E. Burnaford (Eds.), *Images of schoolteachers in the twentieth-century America-paragons, polarities, complexities.* New York: St. Martin's Press.

Finson, K. (2002). Drawing a scientist: What we do and do not know after fifty years of drawings. *School Science and Mathematics, 102*, 335–346.

Katz, P., McGinnis, J. R., Hestness, E., Riedinger, K., Marbach-Ad, G., Dai, A., et al. (2011). Professional identity development of teacher candidates participating in an informal science education internship: A focus on drawings as evidence. *International Journal of Science Education, 33*, 1169–1197.

Mackeracher, D. (1996). *Making sense of adult learning.* Toronto, Ontario, Canada: Culture Concepts Inc.

Plaskoff, J. (2003). Intersubjectivity and community building: Learning to learn organizationally. In M. Easterby-Smith & M. A. Lyles (Eds.), *The Blackwell handbook of organizational learning and knowledge management* (pp. 161–184). Malden, MA: Blackwell Publishing.

Rommetveit, R. (1979). On the architecture of intersubjectivity. In R. Rommetveit & R. B. Blakar (Eds.), *Studies of language, thought and verbal communication.* London: Academic.

Van der Veer, R., & Valsiner, J. (1991). *Understanding Vygotsky: A quest for synthesis.* Cambridge, MA: Blackwell Publishers.

Vygotsky, L. S. (1978). *Mind in society: The development of higher psychological processes.* Cambridge, MA: Harvard University.

Vygotsky, L. S. (1981). The genesis of higher mental functions. In J. V. Wertsch (Ed.), *The concept of activity in soviet psychology* (pp. 144–188). Armonk, NY: Sharpe.

Vygotsky, L. S. (1986). *Thought and language.* Cambridge, MA: MIT Press.

Vygotsky, L. S. (1994). The problem of cultural development of the child. In R. Van Der Veer & J. Valsiner (Eds.), *The Vygotsky reader* (pp. 57–72). Cambridge, MA: Blackwell.

Weber, S., & Mitchell, C. (1996). Drawing ourselves into teaching: Studying the images that shape and distort teacher education. *Teaching and Teacher Education, 12*(3), 303–313.

Author Biographies

The Author of the work is Lillie R. Albert

Lillie R. Albert an associate professor at Boston College Lynch School of Education, has a Ph.D. in Curriculum and Instruction from the University of Illinois at Urbana-Champaign. Her research focuses on the influence that sociocultural historic contexts have on learning and development of learners across the lifespan. Her specialization includes the exploration of the relationship between the teaching and learning of mathematics and the use of cultural and communicative tools to develop conceptual understanding of mathematics. She has published her research in leading national and international journals in her field and presented papers at major research conferences. Her other books include *The Decision to Learn* with Monica Digman and *Student Self-Identified Multiple Intelligence Profiles: A Systematic Technique for Successful Mathematical Problem Solving* with Sheila Cutler Sohn.

She has served as the primary investigator and as a research associate for a number of funded projects by the GE Foundation and the National Science Foundation. She is an active member of the National Council of Teachers of Mathematics and the American Educational Research Association. Other activities include serving as an editorial review panelist for several professional journals; working with the National Research Council of the National Academies as an education panelist; participating as a panelist and site PI for TEDS-M Mathematics and Mathematics Pedagogy Scale Anchoring Study, which is carried out under the sponsorship of IEA and directed by researchers at Michigan State University; and collaborating with mathematics education scholars at Seoul National University of Education, Seoul, South Korea to explore government policies in supporting the preparation of mathematics teachers.

In Collaboration with

Danielle Corea graduated from Boston College in 2010 with a bachelor degree in English and Human Development. She worked as a Research Fellow under Prof. Albert for 3 years of her undergraduate study. She later completed a year of post-graduate service through AmeriCorps VISTA in the Center for Faith and

Public Life at Fairfield University in Connecticut. Currently, she is the program coordinator for the Center. She focuses her work on service learning and the Jesuit Universities Humanitarian Action Network (JUHAN).

Vittoria Macadino graduated from Boston College in 2010 with a Bachelor's Degree in Secondary Education and Mathematics. She was the recipient of the Albert A. Bennett Award for outstanding academic achievement in mathematics. In 2011, Ms. Macadino completed a Master's Degree in Curriculum and Instruction through the Fifth Year Program at Boston College. For all 5 years of her under-graduate and graduate studies, she worked closely under Prof. Lillie Albert as a teaching assistant and Research Fellow contributing to the research, analysis, writing, and editing of many book chapters and articles on mathematics education. Most recently, she collaborated with Dr. Albert and graduate student Karen Terrell on an article entitled *The Mathematics Excellence Partnership: Developing and Sustaining Professional Learning Communities,* currently *in press* in the *Journal of Mathematics Education Leadership.* For the past 2 years, Ms. Macadino has also worked as a mathematics tutor in the Options Through Education Summer Transitional Program, helping to prepare a diverse group of students for entry into Boston College. She is currently teaching mathematics at Newton North High School in Newton, Massachusetts.

Index

L.R. Albert, *Rhetorical Ways of Thinking: Vygotskian Theory and Mathematical Learning*, DOI 10.1007/978-94-007-4065-5,
© Springer Science+Business Media Dordrecht 2012

15798443R00063

Printed in Great Britain
by Amazon